Impacts of Future Weather and Climate Extremes on United States Infrastructure

Assessing and Prioritizing Adaptation Actions

Task Committee on Future Weather and Climate Extremes

Mari R. Tye, Ph.D., CEng
Jason P. Giovannettone, Ph.D., P.E.

Published by the American Society of Civil Engineers

Cataloging in Publication Data on file with the Library of Congress

Published by American Society of Civil Engineers
1801 Alexander Bell Drive
Reston, Virginia 20191-4382
www.asce.org/bookstore | ascelibrary.org

Errata: **Errata, if any, can be found at https://doi.org/10.1061/9780784415863.**

ISBN 978-0-7844-1586-3 (print)
ISBN 978-0-7844-8372-5 (PDF)

Manufactured in the United States of America.

27 26 25 24 23   2 3 4 5 6

Contents

Contributing Authors

Amir AghaKouchak, Ph.D., P.E., M.ASCE, University of California, Irvine, Professor, Center for Hydrometeorology and Remote Sensing

R. Edward Beighley, Ph.D., M.ASCE, Northeastern University, Professor and Associate Chair for Undergraduate Studies, Civil and Environmental Engineering

William J. Capehart, Ph.D., South Dakota School of Mines and Technology, Associate Professor, Department of Civil and Environmental Engineering

Noah J. Fehrenbacher, P.E., S.E., Wiss, Janney, Elstner Associates, Inc., Senior Associate

Robert E. Fields, P.E., Senior Environmental Engineer

Joshua Huang, A.M.ASCE, Technical University of Munich, Research Associate

Laurna Kaatz, Denver Water, Climate Program Director, External Affairs Division

Ning Lin, Ph.D., A.M.ASCE., Princeton University, Associate Professor, Department of Civil and Environmental Engineering

Dagmar Llewellyn, US Department of the Interior Bureau of Reclamation, Civil Engineer (Hydrologic), Albuquerque Area Office (opinions expressed represent professional opinions of the coauthor, not official positions of the government)

Karen MacClune, Ph.D., ISET-International, Climate Change and Hydrological Scientist

J. Rolf Olsen, Ph.D., A.M.ASCE, US Army Corps of Engineers, Senior Technical Lead, Institute for Water Resources (the views expressed are those of the coauthor and do not necessarily represent those of the US Army Corps of Engineers)

Ariane O. Pinson, Ph.D., US Army Corps of Engineers, Climate Science Specialist, Albuquerque District (the views expressed are those of the coauthor and do not necessarily represent those of the US Army Corps of Engineers)

Ting Shi, P.E., M.ASCE, US Federal Energy Regulatory Commission, Civil Engineer (opinions expressed represent professional opinions of the coauthor, not official positions of the government)

Farshid Vahedifard, Ph.D., P.E., F.ASCE, Mississippi State University, CEE Advisory Board Endowed Professor and Associate Professor, Department of Civil and Environmental Engineering

Acknowledgments

The Committee on Adaptation to a Changing Climate (CACC) subcommittee on Future Weather and Climate Extremes is grateful to Ana Barros, Ellen Douglas, Auroop Ganguly, Ben Lord, and Constantine Samaras for their contributions in planning the outline of this book. To ensure that the most accurate and complete information was included in this book, CACC sought reviews from a number of individuals, both internal and external, to ASCE. The insights and suggestions of reviewers from the American Meteorological Society (Michael Dettinger, Rob Cifelli, Kathleen Miller, Francisco Munoz, Tye Parzybok, Julie Vano, and Anna Wilson) and several independent reviewers (Dave Dzombak, Yuchuan Lai, Upmanu Lall, Anne Stoner, Dan Walker, and Daniel B. Wright) substantially added to the content and clarity of this document.

Executive Summary

The condition of a nation's infrastructure and its fragility or resilience to extreme weather and climate events are the foundation of its overall economic strength and position in the world. The systems described in this book are all critical, without which higher-order activities such as markets, education, and health care are impossible. To maintain and build our society, we not only need to maintain current infrastructure functionality but also improve on it, taking into account not only the challenges of today but also future concerns such as climate change, population growth, and urban development. However, ecological resilience indicates that transformation is likely following a major loss of system functionality. Transformation poses a considerable opportunity for resilience but also a threat if designs do not consider emergent risks and technologies associated with alternative system states. The current book focuses on both the vulnerability of the infrastructure of the United States to current weather and climate extremes and the fragility of these systems in the face of climate change.

Most infrastructure is susceptible to the impacts of a changing climate as it is continuously exposed to fluctuating weather conditions. The damage and failure that occurs when exposed to conditions outside its designed range has cascading impacts on community resilience and economies. The Fourth National Climate Assessment report and Intergovernmental Panel on Climate Change (IPCC) Assessment and Special Reports state that the rate of global warming has substantially outpaced natural climate variability and that the effects are evident not only in higher global mean temperatures but also in the frequency and intensity of disruptive events. Principal among the meteorological and hydrological factors that have, and will further, affect civil engineering infrastructure are precipitation extremes, temperature extremes, sea-level rise, and tropical cyclones. The co-occurrence or sequential occurrence of any of these can exacerbate impacts, such as postwildfire flooding or compounded storm surge and higher sea levels. A slew of natural hazards have occurred during recent years that have translated into catastrophic events within the United States. These events have brought consequences such as major infrastructure failures and cascading consequences for the communities involved. With the potential for an increased frequency of disruptive weather and climate events, in addition to increased exposure and interconnectedness as urban and coastal populations grow, it is important to ascertain the resilience of US infrastructure to future natural hazards. Achieving resilience under new and untested climate conditions requires a paradigm shift in design and planning, including preventing infrastructure development in areas with high exposure, and accommodating the potential for failure in design.

US public infrastructure is in an increasingly deleterious state, and there is currently a funding shortfall of approximately $27 trillion over the next 10 years to bring critical infrastructure up to a state of good repair. In addition, much of the existing infrastructure of the United States was designed for the climatic conditions of the twentieth century, with most current engineering minimum design standards not updated to account for recent and future climate change impacts. Rising sea levels, increasing precipitation, and heat extremes are already affecting infrastructure resilience and durability, with increasingly adverse effects expected in many regions. These changes, in addition to the current state of disrepair, have critical ramifications for safety, environmental sustainability, economic vitality and mobility, and system reliability, particularly for vulnerable populations and urban infrastructure. Although funding for infrastructure investment is more strained than ever, additional funding on top of the $27 trillion mentioned is needed to enhance infrastructure resilience to climate change. It has been estimated that adequate investment now could prevent almost 70% of losses from future weather and climate extremes. The magnitude of adequate funding makes it difficult for any single organization to bear the added costs of designing and building more climate-resilient infrastructure. Justifications to apportion the necessary funds are also complicated because of the fact that methods for evaluating the long-term financial benefits of resilient design are relatively nascent. Addressing this financing gap is a challenge that requires a multipronged approach including insurance, taxes, dedicated financing streams, public–private partnerships, and collaboration between sectors and different actors.

Drawing from the 16 categories used in the ASCE's infrastructure report card, this book focuses on the most critical sectors to support a functional society. Critical sectors are aggregated into five overarching infrastructure sectors: energy, transportation (roads and bridges, rail, transit, aviation), drinking water and wastewater, water storage and flood protection, and navigation (including ports and harbors). Each sector is reviewed with respect to the potential impacts of climate change, current sector resilience, adaptation readiness, and dependency on or contribution to other sectors. The authors' assessments for each sector are summarized in the following.

- Energy (transmission, storage, and distribution): This sector is considered ill-prepared to cope with the effects of climate change. In particular, electricity generation will be affected by extreme temperatures, precipitation, and winds and will suffer decreased efficiency as temperatures rise. Demand for electricity is expected to simultaneously increase significantly in the future because of increased cooling demands, increased population, and a growing number of electric vehicles on the road. Because of its interdependency with all other critical infrastructure, energy infrastructure is characterized as extremely critical with regard to its need for adaptation actions.

- Transportation (roads, bridges, transit, and aviation): Coupled with the aging and deteriorating infrastructure network, its location in flat low-lying areas, in general, heightens transportation exposure and sensitivity to disruptive weather and changing climatic conditions. As with the energy sector, transportation is highly interlinked with other sectors, often providing critical support routes for evacuation and reconstruction activities. Vulnerability studies to date have focused on individual assets rather than the full network and the associated costs of replacement, but not on indirect costs from avoided disruptions. The lack of redundancy in all transportation forms makes this sector particularly vulnerable to disruptive weather events that exceed design conditions.

- Water and wastewater: Drinking and wastewater systems are aging and increasing in fragility. Their location near rivers and coasts makes them highly exposed to flooding from fluvial, pluvial, and tidal sources, and the attendant effects on water quality. Although the current risk of failure is moderate, because of built-in adaptive capacity, this will increase in the future without any changes to current management practices. Available potable water in regions where reservoir systems were built to leverage seasonal snowmelt will also likely be affected by changes in precipitation cycles. Given the fundamental importance of water and wastewater, future changes in the climate, together with increased demand and changes in consumption (e.g., arising from changes in population and industrial or agricultural demands), will result in severe impacts in the event of a loss.

- Flood protection infrastructure (levees, flood barriers, dams, and reservoirs): Although dam and reservoir operations are fairly responsive to precipitation variability under current river management practice, water supply is sensitive to future changes in precipitation patterns and the adequacy of storage. Projected changes in precipitation and temperature are likely to increase reservoir sedimentation and decrease water quality, as well as compromise dam structural integrity. The ramifications of dam, levee, or other flood barrier failure are considerable because of the levels of population and infrastructure protected. With many dams and levees having exceeded their design life, failures may occur unless monitoring is increased and safety and maintenance standards are revised.

- Navigation (inland waterways, port, and harbors): Inland navigation is currently more vulnerable to lock failure than to climate extremes. However, competition for water supply during periods of drought may increase the frequency of extended low-flow-induced closures. The impacts of port closures are largely economic and experienced by shippers. However, more frequent or extended closures could exceed the spare capacity in the transportation network with far-reaching consequences on food imports, for example. Port infrastructure is at high risk of closures from coastal storms and sea-level rise. In particular, the supporting infrastructure, such as energy supplies and land transportation, will affect ports' ability to function in the future.

In addition to considering sector vulnerability from the perspective of the vulnerability to current conditions, actions also need to be taken now to reduce future fragility, particularly where those actions involve considerable change to the current modus operandi. Maintaining the efficient operation of drinking water and wastewater systems is critical to the operation of human settlements. Even brief unplanned interruptions to water supply, in addition to energy supply, can compromise emergency services and businesses. In combination with addressing the aging distribution network to reduce leakage, consideration needs to be given to longer-term structural actions to secure future water supplies and wastewater treatment needs.

The ever-increasing interdependency between sectors is also contributing to increased vulnerability and system fragility. There is a need to take a systems-wide approach to risk assessment and interventions and to shift from the traditional fail-safe design paradigm to one that promotes *graceful failure* and supports less tangible benefits such as ecosystem services. Should infrastructure deterioration persist, or investment and construction continue to focus on traditional design solutions, there is a very real risk of increasing catastrophic failures as climate change impacts unfold. However, the aging infrastructure also presents an opportunity to make broad changes to infrastructure system designs that adopt novel and emergent technologies, consider future climate risks, and increase redundancy or reliance on single systems.

Estimating the costs associated with different adaptation scenarios and the avoided costs of failure requires greater research that is beyond the scope of this book. Progress is being made to assess the likely costs of failure, with many agreeing that taking pre-emptive action now will be less of a financial burden than recovery at a later stage. However, direct project finances are not the only consideration, the economic (e.g., job loss or creation), social, and environmental costs and benefits must also be examined. In prioritizing the best allocation of limited project funds, pertinent questions and ranking criteria are required, such as

- Where is the greatest sensitivity, exposure, or impact to disruptive events?
- Where can the greatest risk reduction be achieved?
- Are there interdependencies with the potential to cause cascading failures?
- Where is the greatest infrastructure investment gap?
- What are the social, environmental, and economic costs and benefits?

Applying the combined experience of an interdisciplinary team of climate scientists and engineers to a tentative prioritization of the greatest need for investment in US infrastructure is presented. This is a qualitative ranking that will differ considerably by region, jurisdiction (e.g., state, county, or city), and the communities involved. Although all infrastructure is in a poor standard and exhibiting deterioration, with the infrastructure score of D+, the impacts of failure and continued deterioration in some sectors will have greater consequences than from others. The interdependence between sectors and societal reliance on the energy, highways (roads and bridge), and flood infrastructure will result in widespread disruptions and prolonged economic, social, and environmental consequences, making these a priority at a national level.

An interdisciplinary approach involving policymakers, economists, and other experts will also be needed to achieve the considerable changes needed in specific planning, design, construction, maintenance, and decommissioning processes. Collaborative research is needed to understand how climate change will impact specific locations, regions, and assets over different time scales. Improved two-way communication will be essential to address uncertainties in climate modeling and to integrate these considerations into engineering and design. Finally, the policies and laws that inform infrastructure decision-making and investment require a thorough review to integrate climate change considerations into decision-making and design processes more effectively.

The purpose of this book is to build on the fundamental knowledge presented by previous ASCE publications by reviewing the current status of research on the impacts of climate change on infrastructure. It summarizes the likely changes in various extreme meteorological and hydrological events and assesses the vulnerabilities of critical sectors, and their collective interdependencies, to the negative impacts of said events. In addition, frameworks that decision-makers can use to prioritize limited budgetary resources for adaptation efforts are reviewed. Although the immediate aftermath of failure, or its avoidance, immediately leaps to mind, this book considers both acute vulnerabilities and responses (e.g., immediate consequences from a hurricane) and the chronic vulnerability to climate change. Not only is there a need to improve the capacity for emergency response and speed up the recovery process, it is also necessary to consider the major capital investments that could avoid interruptions of essential services. Recent disasters have emphasized that the loss of energy, transportation, and telecommunications affects day-to-day life and has short- and long-term economic consequences. However, over a longer time frame, major disruptions, such as a loss of water supply due to increased droughts, reduced groundwater, or less snowpack, will have irretrievable consequences that can be avoided or mitigated by taking actions now and by properly prioritizing these actions.

CHAPTER 1

Introduction

1.1 CURRENT NEEDS

The strength of the US infrastructure is the foundation of our economic strength and position in the world and depends on various critical systems for its maintenance and longevity; Figure 1-1 illustrates the hierarchy of systems needed to assure a robust and thriving society. The cornerstone underlying everything else is an interconnected network of healthy, functioning ecosystems, providing clean air and water, food, and raw materials. In addition to water and food, clean and readily accessible energy comes next, underpinning life and economic activity across the country. A dense transportation network, including roads, trains, airplanes, and boats, is fundamental to moving goods and services and is particularly critical to

Figure 1-1. Hierarchy of systems in terms of their criticality for human society to survive and thrive.

Source: ISET-International (2020).

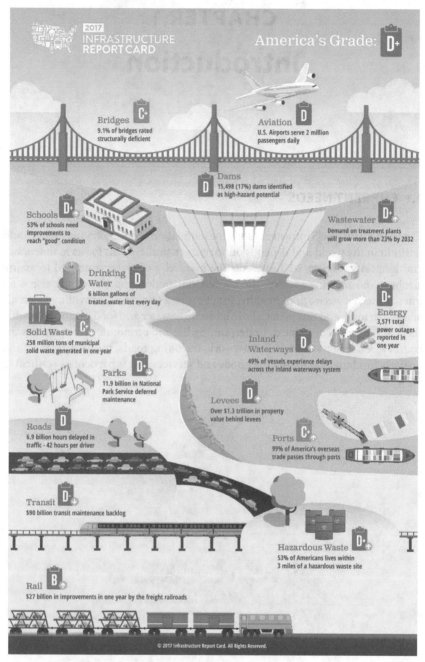

Figure 1-2. ASCE infrastructure report card; an assessment of the condition of the nation's infrastructure and the needs to raise the grades.

Source: ASCE (2017).

the operation of urban centers where the need for goods and services outpaces the ability of the local environment to provide them. Safe and secure housing is also found at this level. The final tier of core services includes a robust communications network, critical to economic endeavors, and accessibility to an effective sanitation system, which is essential for maintaining public health.

The systems described in this book—energy, transportation, drinking water and wastewater, water storage and flood protection, and navigation/ports and harbors—are all critical to the functioning of these core systems. Without these systems, higher-order activities such as markets, education, health care, and all the other elements of a developed society—shown in the upper half of the triangle—are impossible. To maintain and build our society, we need to not only maintain the current functioning of these foundational systems but also improve them, taking into account both the challenges of today and those posed by future population growth, urban development, and changes in our environment and climate.

ASCE provided an assessment of the current condition and needs of 16 major infrastructure categories within the United States and assigned each a grade from A to F; the grades as of 2017 are shown in Figure 1-2. As can be seen in Figure 1-2, the United States has a rapidly emerging crisis associated with aging infrastructure, exemplified by its D+ cumulative infrastructure grade point average (ASCE 2017) and the fact that only four individual sectors received a grade better than a D+. Infrastructure is defined as "the basic physical and organizational structures and facilities needed for the operation of a society or enterprise" (OED Online 2014). The United States spends approximately 2% of its gross domestic product (GDP) on infrastructure, estimated at $36.8 billion annually. Infrastructure also enables income and growth: Globally, nearly 40% of the GDP was attributed to this sector, $5.4 trillion in the United States (Arcadis 2016). The infrastructure deficiencies identified led to 3,571 total power outages in 1 year, 6 billion gallons of water lost daily, 49% of inland vessels experiencing delays within American waterways, and billions of hours of traffic delays (ASCE 2017). However, this grading scheme does not take into account the resilience or fragility of a said infrastructure to future perturbations caused by climate change, population growth, and urban development.

1.2 CLARIFICATION OF TERMINOLOGY

Throughout this book, certain definitions have been selected from the literature and are included in the Glossary. Given the variety of definitions adopted by different sectors, specific terms are called out and explained in greater detail in the following paragraphs.

1.2.1 Adaptive Capacity

Intergovernmental Panel for Climate Change (IPCC 2014) defines adaptive capacity as "The ability of systems, institutions, humans and other organisms to adjust to potential damage, to take advantage of opportunities, or to respond to consequences." The flexibility for adjusting to change depends on both human system and policy components. Human system components range from adjusting inspection and maintenance intervals (e.g., of trash screens on culverts), through upgrading road surface materials as they reach the end of life, to designing enhanced foundations to facilitate elevation increases in sea walls at a later stage. Another aspect of flexible adaptation examines the infrastructure operating conditions. For instance, heating, ventilation, and cooling (HVAC) systems in a building may be operated within a range of design conditions. Although the HVAC systems may not be directly affected by an increase in exceptionally hot or cold days, the additional accumulated burden in relation to the designed number of exceedances may reduce the overall efficiency of the system or its operating life.

Flexible adaptation, also known as dynamic adaptation planning (Rosenzweig et al. 2011, Yohe and Leichenko 2010), has been practiced in some guise in the field of ecology for more than 40 years (Holling 1973), with a more recent provision for decision-making under conditions of deep uncertainty (Hallegatte et al. 2012). Incorporating policy changes within this dynamic provides a more regularized framework to ensure that trigger points are identified in a timely manner; Haasnoot et al. (2013) refer to this combination as dynamic adaptive policy pathways comprising adaptive policymaking and adaptation pathways. Adaptive policymaking identifies a basic plan for adaptation activities and subsequent contingency plans to react flexibly as new information becomes available. Adaptation pathways develop a sequence of actions over time, specifically identifying when certain actions may no longer be adequate.

In effect, adaptation pathways require the decision-maker to postulate and solve for a sequence of investments or actions, contingent on events that may occur, rather than solving for the best investment or action at present. In the case of climate adaptation, the motivation is to avoid the potential damage and rebuilding costs arising from future disruptive weather or climate events without compromising future decisions. For instance, a decision as to whether a design may be enhanced so that the structure can be altered more readily at a later date may be made. On a different level, the adaptation may center around investment in a technology, where cheaper and more efficient alternatives may emerge and render the original technology obsolete. Then, the adaptation pathway seeks to minimize the risks of not realizing the projected return on investment when users switch to the new technology or of being prevented by political factors from transitioning to the newer, more efficient systems.

1.2.2 Disruptive Events

McPhillips et al. (2018) suggest that the confusion, occasionally leading to inaction, around disruptive events lies in the lack of clarity about what comprises an extreme. The authors found that there is a stark difference between how different disciplines define an extreme, particularly between scientists, who focus on the impacts, and engineers, who focus on event magnitude and frequency. Scientists within any of the Earth science disciplines define an extreme, which is referred to as a hazard, based on its potential negative impact on people, infrastructure, or the environment. An extreme event is referred to as a disturbance within the field of ecology and is defined based on the magnitude by which ecological structure and function are enhanced or inhibited by the disturbance. The social conditions that are either impacted by, or themselves influence, extreme events are the primary focus of hazard analysis within the social sciences.

In contrast to the primary focus of the aforementioned scientific disciplines, engineers typically define extremes based on specific return periods (e.g., 100 year discharge), which are required in the selection of design standards; the magnitude for a specific return period is computed using long-term event data (e.g., streamflow data records). Regardless of the specific definition used, McPhillips et al. (2018) found that almost a quarter of the papers they reviewed conflated definitions of extreme impact and extreme events. The general nature of the current book, therefore, encompasses both those events that are rare or extreme and those that can have high consequences. It is also worth noting that rarity is subjective and is dependent on the discipline: Although IPCC (2012) identifies rare events as those "rarer than the 10th or 90th percentiles of the observed probability density function," civil engineers may consider rarity to represent an acceptable probability of a dam failing (ASCE CACC 2018). In the interest of clarity, disruptive events are composed of severe, rare, extreme, and high-impact events (Stephenson 2008).

1.2.3 Potential Impact (i.e., Exposure and Sensitivity)

As illustrated in Figure 1-3, the potential impact of various aspects of climate change on a particular sector can be measured by considering both the magnitude

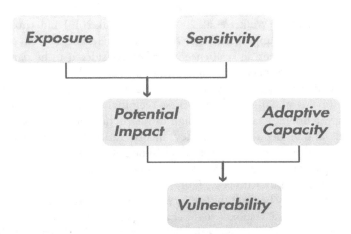

Figure 1-3. Key components of vulnerability, illustrating the relationship among exposure, sensitivity, and adaptive capacity.
Source: Glick et al. (2011).

of its projected response to (i.e., sensitivity) and the extent to which the sector is likely to be affected (i.e., exposure) by a particular manifestation of climate change. The likelihood of and the magnitude of the response of a sector to elements of climate change characterize its sensitivity, which may depend on factors such as location, the ability of infrastructure to accommodate further change, dependence on hydrological or meteorological changes (i.e., snowmelt and precipitation), and its dependence on the full functionality of other sectors (i.e., its level of interdependency), and vice versa.

The total potential impact of a climate change stressor on a particular sector also depends on its level of exposure to that stressor. Even if a sector is inherently highly sensitive to a stressor but experiences minimal exposure, the potential impact will be minimal. For example, the transportation sector is extremely sensitive to flooding, whether it be caused by an extreme precipitation or riverine flood event or a storm surge from a coastal event exacerbated by sea-level rise. The total potential impact on the sector in a particular region or location will be highly dependent on the physical characteristics that define the location of specific infrastructure, such as roads and bridges. If a particular road is located at a relatively high elevation and a sufficient distance from any river or coastline, its level of exposure will be low. Thus, even though roads, in general, are considered highly sensitive to flooding, due to the limited exposure the potential impact on the aforementioned road would be low.

1.2.4 Resilience

Consider the definition of civil engineering as "the art of directing the great sources of power in nature for the use and convenience of man" (Tredgold 1828). By changing the emphasis instead to working *with* nature, the tradition of combatting the forces of disruptive weather and climate events with resistive designs moves toward achieving resilience in a way that blends ecological and engineering theories (Walker et al. 2004, Linkov et al. 2013) and, thus, is more appropriate for a changing climate. As noted by Holling (1996), ecological resilience and engineering resilience are often considered to be paradoxical. The former embraces persistent change and unpredictability, whereas the latter centers on efficiency, constancy, and predictability. However, a *graceful failure* approach (Tye et al. 2014) combines these seemingly contradictory notions by accepting the unpredictability to design infrastructure that has flexible design thresholds and embedded redundancy. Incorporating failure does not mean that lifelines such as emergency services will be put at risk, rather the intention is to anticipate and mitigate against the potential consequences should failure occur. As identified by Kim et al. (2017), *fail-safe* designs may only exacerbate risks that are already being amplified by a changing climate. Thus, resilience is a dynamic property that responds to the changing conditions presented by climate and anthropogenic developments through providing redundancy and alternate means of operation. Specifically, designing an infrastructure component or a system for failure ensures that essential functions are maintained, embraces the uncertainties of current and future climate conditions, and directs the greatest impacts to where they can be absorbed.

1.2.5 Risk

Drawing on the IPCC definitions, risk is a continuum that depends on exposure and vulnerability and can vary for different hazardous events, locations, and people affected. Although the term *hazard* has in the past been considered synonymous with risk, current definitions acknowledge that hazard is a component of risk, not the actual risk (Cardona et al. 2012). The IPCC Summary for Policy Makers (2012) further clarifies that *disaster risk* occurs at the intersection of exposure, disruptive weather and climate events, and vulnerability and can be tempered or exacerbated by human interventions and adaptation and external forces such as climate (Figure 1-4). Risk can be calculated as the probability of a hazard occurring multiplied by impacts if disruptive events occur, giving the potential for adverse consequences to arise.

1.2.6 Vulnerability

Bakkensen et al. (2016) noted that many attempts to quantify vulnerability are similar in nature to definitions of resilience, despite the differences in their underlying theoretical constructs. IPCC (2014) defines vulnerability as "The propensity or predisposition to be adversely affected. Vulnerability encompasses a variety of concepts and elements including sensitivity or susceptibility to harm and lack of capacity to cope and adapt." Glick et al. (2011) aimed to quantify vulnerability by defining it as "a function of the sensitivity of a particular system to climate changes, its exposure to those changes, and its capacity to adapt to those changes," as illustrated in Figure 1-3.

In contrast, the risk modeling industry defines vulnerability as a distinct concept that is independent of exposure. The computer industry considers vulnerability as the properties of a system that make compromise possible and, hence, pose a risk (Otwell and Aldridge 1989). In a similar vein, engineering disciplines consider vulnerability as the relationship between a measure of environmental damage potential (e.g., flood volume or depth) and the degree of loss (e.g., repair cost). This relationship is also described variously as fragility or damage for different perils (MMC 2018). Fragility could, therefore, be considered as the antonym of resilience or the "quality of being easily broken or damaged" (Kennedy et al. 1980) and the residual risk of failure.

IPCC (2014) definition of vulnerability has been adopted throughout this study to distinguish between these different concepts, whereas fragility is used for the engineering definition. That is, fragility represents the continued probability of failure given the system(s)' lack of capacity to absorb and recover from persistent threats or changing conditions and is the antonym of resilience.

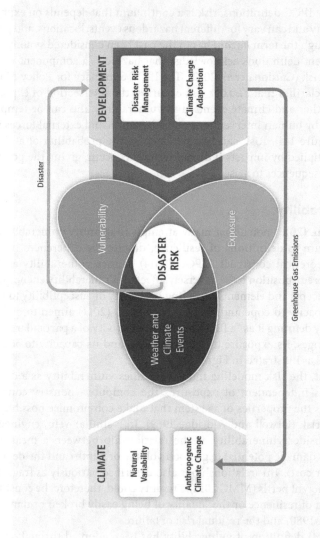

Figure 1-4. Illustration of the core concept of disaster risk and its dependence on exposure, vulnerability, human interventions, and climate.
Source: SREX Fig. SPM.1 in IPCC (2012).

1.3 FUTURE CONCERNS

Society as a whole is increasingly complex and, as a result of the interdependence between infrastructure systems and a lack of redundancy, vulnerable to disruptive weather and climate events (ASCE CACC 2018, Chang 2016). Increases in population and urban development in highly exposed hazard-prone areas such as the coastal zone amplify this vulnerability (Hinkel et al. 2014, Swiss Re 2014). Severe storms, together with associated disruptive weather events such as tornadoes, hail, and heavy rain and wind, impact lives and property globally, often resulting in direct losses of billions of dollars per event (Munich Re 2018), with further billions lost in hard-to-quantify long-term and far-field impacts such as business interruptions or health-related impacts. Furthermore, an increasing interdependence between infrastructure systems, in combination with the effects of climate change and population growth, all contribute to an increasing sensitivity and exposure (i.e., overall vulnerability) and a greater probability of catastrophic failures and overall increasing system fragility (Chang 2016).

The exposure to rare and disruptive events has grown over the last 40 years through a combination of increasingly frequent disruptive weather events and population growth in exposed areas (Figure 1-5), and this is expected to continue to increase in the future as both these factors continue to increase (Knutson et al. 2010, Seneviratne et al. 2012). For example, in 2010, 123.3 million people

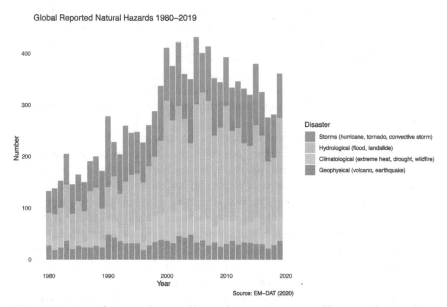

Figure 1-5. Annual reported natural hazard events categorized by type. The numbers include both weather and nonweather disasters.

Source: EM-DAT (2020).

lived in US coastal shoreline counties (NOAA 2013); most of these counties and their infrastructure are also at high risk from disruptive weather events such as hurricanes. As a result, many of the aforementioned deficiencies identified in the ASCE report card, among others, will be exacerbated by a higher frequency and intensity of future extreme meteorological and hydrological events, population shifts, and development in highly vulnerable regions.

In recent years, the United States has experienced several disruptive weather events that have resulted in catastrophic losses (both financial and human). From shorter-duration events such as Hurricanes Sandy (2012) and Harvey (2016) to events that played out over a longer period of time such as the California droughts, wildfires, and floods (2012 to 2018), each had direct impacts on infrastructure and cascading consequences for the communities involved. Although the indirect costs of public infrastructure failures are hard to quantify, this burden is largely borne by private entities and far eclipses the cost of replacing the original infrastructure. When Hurricane Maria made landfall over Puerto Rico in September 2017, parts of the island were still recovering from Hurricane Irma's passage, which occurred almost 2 weeks prior. Remnants of the lingering effects of older storms, such as Hurricanes Katrina (2005) and Sandy (2012) (Government of Puerto Rico 2018), were also still present because of delayed recovery efforts. The combination of high sensitivity and exposure led to complete failure of the electrical power system and associated failure of the water supply system, both of which had severe health impacts as people were unable to access critical emergency and ongoing medical services.

Although each of the aforementioned weather events culminated in disaster, the events themselves did not necessarily presage disaster. Their transformation into disaster arose from the combination of heightened exposure owing to climate change and increased sensitivity of the local populations and infrastructure to those hazards. The cascade of impacts and prolonged recovery time serve to highlight how the interconnectedness of modern society accentuates sensitivity, and in effect vulnerability, to the long-term consequences of disruptive weather events.

Focusing only on the most disruptive weather events risks minimizing the cumulative financial effects of repeated exposure to less severe but more frequent, nondisastrous events. For instance, *nuisance* or *sunny-day* flooding in coastal regions is becoming more frequent (Ezer and Atkinson 2014, Moftakhari et al. 2017, Sweet and Park 2014), particularly where low-lying land is highly urbanized and reliant on aging surface water drainage systems (Wahl et al. 2015). This leads to more than 100 million vehicle hours of delay annually along the US East Coast (Jacobs et al. 2018a). Given increasing urbanization and reduced permeability, inland areas are not exempt from this problem (Moftakhari et al. 2018, Schröer and Tye 2019), with estimated cumulative damage in the United States from summertime convective storms during 2016 almost equaling that from Hurricane Katrina (Munich Re 2018).

Infrastructure failures from disruptive weather and climate, and from more common *nuisance* events, can have wide-reaching consequences extending away from the site of the original disaster and often for a considerable duration after immediate failure. Of the annual losses from disastrous weather and climate events, research by Swiss Re suggests that up to 68% of losses can be avoided with an appropriate adaptation of infrastructure (Franzke 2017, Swiss Re 2014). However, meeting the global demand to improve infrastructure resilience to weather and climate extremes over the next 10 years will require investments on the order of $1 trillion per year (Bhattacharya et al. 2016). The lack of redundancy in many infrastructure systems makes them particularly vulnerable to changes in extreme weather exposure that will be brought on by climate change (Mattsson and Jenelius 2015). As illustrated in Figure 1-6, a flooded energy substation could affect water and wastewater distribution, telecommunications, and building operations (Kunreuther et al. 2009) with escalating effects if repairs cannot be achieved because of flooded access routes (Sudhalkar et al. 2017). For example, the subsequent floods and landslides from the aforementioned disruptive weather events in Puerto Rico caused an extended disruption of inland transportation. Because of their dependence on a functional transportation network, the restoration of power and other critical services was delayed even after critical water supplies and backup generators had been shipped in.

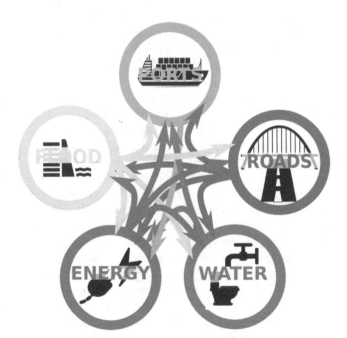

Figure 1-6. Interdependencies between different infrastructure sectors (flood protection, ports and waterways, transportation, energy, and water), highlighting the importance of energy and transportation to the resilience of other sectors.
Source: Infographic B-2 in Sudhalkar et al. (2017).

1.4 ADAPTATION OPTIONS AND PRIORITIZATION

Implementing adaptation options for climate change into civil engineering design is complicated by the traditional approaches premised on empirically derived probabilities of risk to characterize functionality, durability, and safety (ASCE CACC 2018). Given the current state of infrastructure, their increasing levels of interdependence, competing financial demands, and anthropogenic stressors, the question arises, to paraphrase Fischhoff et al. (1978), "how resilient is resilient enough?" Or, how do we determine which infrastructure components are most vulnerable to disruptive weather and whose failure would have the greatest societal and economic impacts, or have the greatest fragility. The US Federal Interagency Climate Change Adaptation Task Force (FICCATF 2011) guiding principles included the key step of prioritizing critical infrastructure, people, and places that are most vulnerable to the effects of climate change. However, as demonstrated by subsequent national and international adaptation and resilience activities (Bles et al. 2015, 2016; Choate et al. 2014; CSIWG 2018; Filosa et al. 2018; ICE 2011), this is not only data intensive and time-consuming but is often limited by being considered from a single-sector point of view. In addition, it could be argued that the current lack of resilience stems from the "single discipline, single event, single network" component approach to design (Markolf et al. 2018). That is, a bridge may be designed with the approaching highway in mind, but its function to convey utility cables and pipes or integration into the broader transportation network may be disregarded. Yet, decision-makers are often required to balance the needs of several sectors, but they lack the information to do this adequately.

In this vein, the US Geological Survey Multi Hazards Demonstration Project ARkStorm scenario considered the infrastructural, economic, and societal impacts in California from an extreme weather event (Porter et al. 2011). The scenario assessed the relative roles of immediate infrastructure impacts against long-term disruptions across a wide range of sectors, including losses of building function, lifeline services, and general productivity (Wing et al. 2016). Although this was focused on California, the results give a flavor of how the effects of climate change on critical infrastructure might be felt in other locations and where efforts need to be concentrated both to mitigate against and adapt to change. However, the scenario still does not provide a conclusive guide to the prioritization of limited funding and urgency of issues across several sectors. Benefit–cost analysis approaches are widely adopted to justify infrastructure investment; however, assessing future avoided costs or likely damage is even more challenging under varying risk conditions. Furthermore, the effects of nuisance flood damage or several consecutive higher impact events are not comparable and do not have a linear cost and damage relationship, thus increasing the difficulty of assessing the benefit–cost ratio.

1.5 OVERALL OBJECTIVE

The purpose of this book is to synthesize the available research on individual sector vulnerabilities to disruptive weather and climate and to consider the possible fragility of these systems, given the anticipated changes in climate and adaptation activities. This book builds on the fundamental knowledge presented by previous ASCE publications (ASCE CACC 2015, 2018) to provide guidance to decision-makers in developing their own assessment of urgency for action. In addition to the impacts of weather and climate extremes on individual sectors individually, their many dependencies and interdependencies are also considered.

Most infrastructure is vulnerable to the impacts of a changing climate because it is continuously exposed to outdoor conditions. The damage and failure that occurs when exposed to conditions outside of its designed range (Chinowsky and Arndt 2012) has cascading impacts on community resilience and economies (Jacobs et al. 2018b, Koetse and Rietveld 2009, USGCRP 2018). Furthermore, US infrastructure is in an increasingly deleterious state, and the funding to ensure climate-resilient infrastructure investment is more strained by competing budget needs. Justifications to appropriate the necessary funds are also complicated as methods for evaluating the long-term financial benefits of resilient design are relatively nascent (NASEM 2018a, b).

Any prioritization scheme adopted by decision-makers must first identify the purpose of the prioritization. Is it to focus adaptation efforts across infrastructure types where they will have the most impact, while balancing limited budgets? To focus action where the longer-term consequences require advance action to prevent maladaptation or *lock-in* (Markolf et al. 2018)? Or is to encompass a broad range of uncertainties and reduce the future fragility of the infrastructure element or system? The prioritization must also account for adaptation actions that are in progress or being carried out elsewhere and allow for future flexibility and changes in policy or needs. This book presents a summary of prioritization schemes appropriate for infrastructure projects, together with some examples of how they might be employed.

The intended audience for this book encompasses all those involved in infrastructure planning and design, from federal and statewide officials to those with a more local remit. The synthesis of how changes in disruptive weather events translate into impacts on a specific infrastructure type and the potential cascade of failures stemming from its damage aims to support those producing vulnerability assessments or determining project prioritization. Similarly, the summaries of impacts are a guide for infrastructure designers to accommodate the most likely future weather and climate extremes for their locale while specific design standards are under development. This information can also be used in benefit–cost analysis by focusing only on the most likely consequences of failure and the anticipated scale of costs to repair the infrastructure.

The remainder of this book is set out as follows: Chapter 2 summarizes the state of scientific knowledge on current and future meteorological and hydrological extremes. Chapter 3 outlines how these changes will manifest as impacts to the individual sectors, the status of climate change adaptation efforts to stem these impacts in each sector, potential additional responses to climate change adaptation, and a discussion of the level of interdependency and influence between each sector. These assessments can then be used to inform case-specific prioritization using the sample methods summarized in Chapter 4. The information gleaned from the review of current literature also facilitates a means of determining whether current standards of design are sufficient to accommodate future operating conditions, both extreme weather and day-to-day operations. Therefore, Chapter 5 includes a proposal for necessary future research in addition to more general conclusions stemming from the findings of this book.

References

Arcadis. 2016. "Global built asset wealth index 2015." Accessed June 9, 2019. https://www.arcadis.com/en/global/our-perspectives/global-built-asset-wealth-index/.

ASCE. 2017. *Infrastructure report card. A comprehensive assessment of America's infrastructure*. Reston, VA: ASCE.

ASCE CACC (Committee on Adaptation to a Changing Climate). 2015. *Adapting infrastructure and civil engineering practice to a changing climate*, edited by J. R. Olsen. Reston, VA: ASCE.

ASCE CACC. 2018. *Climate-resilient infrastructure: Adaptive design and risk management*, edited by B. M. Ayyub. ASCE Manuals and Reports on Engineering Practice No. 140. Reston, VA: ASCE.

Bakkensen, L. A., C. Fox-Lent, L. Read, and I. Linkov. 2016. "Validating resilience and vulnerability indices in the context of natural disasters." In Chap. 4 in *IRGC resource guide on resilience*, edited by M.-V. Florin and I. Linkov. Lausanne: EPFL International Risk Governance Center. 23-29.

Bhattacharya, A., J. P. Meltzer, J. Oppenheim, Z. Qureshi, and N. Stern. 2016. *Delivering on sustainable infrastructure for better development and better climate*. Washington, DC: Brookings Institution, New Climate Economy, and Grantham Institute for Climate Change and the Environment.

Bles, T., J. Bessimbinder, M. Chevreuil, P. Danielsson, S. Falemo, and A. Venmans. 2015. "ROADAPT—Roads for today, adapted for tomorrow." Accessed October 17, 2018. http://www.cedr.eu/download/other_public_files/research_programme/call_2012/climate_change/roadapt/ROADAPT_integrating_main_guidelines.pdf.

Bles, T., J. Bessembinder, M. Chevreuil, P. Danielsson, S. Falemo, A. Venmans, et al. 2016. "Climate change risk assessments and adaptation for roads—Results of the ROADAPT project." *Transp. Res. Procedia* 14: 58–67.

Cardona, O. D., M. K. van Aalst, J. Birkmann, M. Fordham, G. McGregor, R. Perez, et al. 2012. "Determinants of risk: Exposure and vulnerability." In *Managing the risks of extreme events and disasters to advance climate change adaptation*, edited by C. B. Field, V. Barros, T. F. Stocker, D. Qin, D. J. Dokken, K. L. Ebi, et al. A Special Report of Working Groups I and II of the Intergovernmental Panel on Climate Change (IPCC), 65–108. Cambridge, UK: Cambridge University Press.

Chang, S. E. 2016. Vol. 1 of *Socioeconomic impacts of infrastructure disruptions*. Oxford, UK: Oxford University Press.

Chinowsky, P., and C. Arndt. 2012. "Climate change and roads: A dynamic stressor-response model." *Rev. Dev. Econ.* 16 (3): 448–462.

Choate, A., C. Evans, B. Rodehorst, R. Saavedro, C. Snow, J. Snyder, et al. 2014. *Impacts of climate change and variability on transportation systems and infrastructure. The gulf coast study, phase 2 screening for vulnerability*. Washington, DC: US Dept. of Transportation and Federal Highway Administration.

CSIWG (Climate-Safe Infrastructure Working Group). 2018. *Paying it forward: The path toward climate-safe infrastructure in California*. Report of the Climate-Safe Infrastructure Working Group to the California State Legislature and the Strategic Growth Council. Sacramento, CA: CSIWG.

EM-DAT (Emergency Database). 2020. *OFDA/CRED international disaster database*. Brussels, Belgium: Université catholique de Louvain.

Ezer, T., and L. P. Atkinson. 2014. "Accelerated flooding along the U.S. East Coast: On the impact of sea-level rise, tides, storms, the Gulf Stream, and the North Atlantic Oscillations." *Earth's Future* 2 (8): 362–382.

FICCATF (Federal Interagency Climate Change Adaptation Task Force). 2011. *Federal actions for a climate resilient nation*. Progress Rep. Washington, DC: FICCATF.

Filosa, G., A. Plovnik, L. Stahl, R. Miller, and D. Pickering. 2018. *Vulnerability assessment and adaptation framework*. 3rd ed. Washington, DC: Federal Highway Administration.

Fischhoff, B., P. Slovic, S. Lichtenstein, S. Read, and B. Combs. 1978. "How safe is safe enough? A psychometric study of attitudes towards technological risks and benefits." *Policy Sci.* 9 (2): 127–152.

Franzke, C. L. E. 2017. "Impacts of a changing climate on economic damages and insurance." *Econ. Disasters Clim. Change* 1 (1): 95–110.

Glick, P., B. A. Stein, and N. A. Edelson. 2011. *Scanning the conservation horizon: A guide to climate change vulnerability assessment*. Washington, DC: National Wildlife Federation.

Government of Puerto Rico. 2018. "Transformation and innovation in the wake of devastation: An economic and disaster recovery plan for Puerto Rico." Accessed December 6, 2018. http://www.p3.pr.gov/assets/pr-draft-recovery-plan-for-comment-july-9-2018.pdf.

Haasnoot, M., J. H. Kwakkel, W. E. Walker, and J. ter Maat. 2013. "Dynamic adaptive policy pathways: A method for crafting robust decisions for a deeply uncertain world." *Global Environ. Change* 23 (2): 485–498.

Hallegatte, S., A. Shah, R. Lempert, C. Brown, and S. Gill. 2012. *Investment decision making under deep uncertainty. Application to climate change*, 1–41. Policy Research Working Paper No. 6193. Washington, DC: World Bank.

Hinkel, J., D. Lincke, A. T. Vafeidis, M. Perrette, R. J. Nicholls, R. S. J. Tol, et al. 2014. "Coastal flood damage and adaptation costs under 21st century sea-level rise." *Proc. Natl. Acad. Sci. U.S.A.* 111 (9): 3292–3297.

Holling, C. S. 1973. "Resilience and stability of ecological systems." *Annu. Rev. Ecol. Syst.* 4 (1): 1–23.

Holling, C. S. 1996. "Engineering resilience versus ecological resilience." In Chap. 3 in *Engineering within ecological constraints*, edited by National Academy of Engineering, 31–44. Washington, DC: National Academies Press.

ICE (Institution of Civil Engineers). 2011. *Built environment/transport briefing sheet: Managing the highway in extreme weather conditions*. London: ICE.

IPCC (Intergovernmental Panel on Climate Change). 2012. "Figure SPM.1 from IPCC, 2012: 'Summary for policymakers.'" In *Managing the risks of extreme events and disasters to advance climate change adaptation*, edited by C. B. Field, V. Barros, T. F. Stocker, D. Qin, D. J. Dokken, K. L. Ebi, et al., 3–21. A Special Report of Working Groups I and II of the Intergovernmental Panel on Climate Change. Cambridge, UK: Cambridge University Press.

IPCC. 2014. "Annex II: Glossary." [Mach, K. J., S. Planton and C. von Stechow (eds.)] In *Climate change 2014: Synthesis report.* Contribution of Working Groups I, II and III to the fifth assessment report of the Intergovernmental Panel on Climate Change, edited by Core Writing Team, R. K. Pachauri, and L. A. Meyer, 117–130. Geneva: IPCC.

ISET-International. 2020. Accessed August 17, 2020. https://www.i-s-e-t.org/what-we-do.

Jacobs, J. M., L. R. Cattaneo, W. Sweet, and T. Mansfield. 2018a. "Recent and future outlooks for nuisance flooding impacts on roadways on the US East Coast." *Transp. Res. Rec.* 2672 (2): 1–10.

Jacobs, J. M., M. Culp, L. Cattaneo, P. Chinowsky, A. Choate, S. DesRoches, et al. 2018b. "Transportation." In Chap. 12 in Vol. 2 of *Impacts, risks, and adaptation in the United States: Fourth National Climate Assessment*, edited by D. R. Reidmiller, C. W. Avery, D. R. Easterling, K. E. Kunkel, K. L. M. Lewis, T. K. Maycock, et al., 479–511. Washington, DC: US Global Change Research Program.

Kennedy, R. P., C. A. Cornell, R. D. Campbell, S. Kaplan, and H. F. Perla. 1980. "Probabilistic seismic safety study of an existing nuclear power plant." *Nucl. Eng. Des.* 59 (2): 315–338. https://doi.org/10.1016/0029-5493(80)90203-4.

Kim, Y., D. A. Eisenberg, E. N. Bondank, M. V. Chester, G. Mascaro, and B. S. Underwood. 2017. "Fail-safe and safe-to-fail adaptation: Decision-making for urban flooding under climate change." *Clim. Change* 145 (3–4): 397–412.

Knutson, T. R., J. L. L. McBride, J. Chan, K. Emanuel, G. Holland, C. Landsea, et al. 2010. "Tropical cyclones and climate change." *Nat. Geosci.* 3 (3): 157–163.

Koetse, M. J., and P. Rietveld. 2009. "The impact of climate change and weather on transport: An overview of empirical findings." *Transp. Res. Part D: Transp. Environ.* 14 (3): 205–221.

Kunreuther, H., E. Michel-Kerjan, and N. A. Doherty. 2009. *At war with the weather: Managing large-scale risks in a new era of catastrophes.* Cambridge, MA: MIT Press.

Linkov, I., D. A. Eisenberg, M. E. Bates, D. Chang, M. Convertino, J. H. Allen, et al. 2013. "Measurable resilience for actionable policy." *Environ. Sci. Technol.* 47 (18): 10108–10110.

Markolf, S. A., M. V. Chester, D. A. Eisenberg, D. M. Iwaniec, C. I. Davidson, R. Zimmerman, et al. 2018. "Interdependent infrastructure as linked social, ecological, and technological systems (SETSs) to address lock-in and enhance resilience." *Earth's Future* 6 (12): 1638–1659.

Mattsson, L.-G., and E. Jenelius. 2015. "Vulnerability and resilience of transport systems—A discussion of recent research." *Transp. Res. Part A: Policy Pract.* 81: 16–34.

McPhillips, L. E., H. Chang, M. V. Chester, Y. Depietri, E. Friedman, N. B. Grimm, et al. 2018. "Defining extreme events: A cross-disciplinary review." *Earth's Future* 6 (3): 441–455.

MMC (Multihazard Mitigation Council). 2018. *Natural hazard mitigation saves: Utilities and transportation infrastructure*, edited by K. Porter, C. Scawthorn, C. Huyck, R. Eguchi, Z. Hu, and P. Schneider. Washington, DC: MMC, National Institute of Building Sciences.

Moftakhari, H. R., A. AghaKouchak, B. F. Sanders, M. Allaire, and R. A. Matthew. 2018. "What is nuisance flooding? Defining and monitoring an emerging challenge." *Water Resour. Res.* 54 (7): 4218–4227.

Moftakhari, H. R., A. AghaKouchak, B. F. Sanders, and R. A. Matthew. 2017. "Cumulative hazard: The case of nuisance flooding." *Earth's Future* 5 (2): 214–223.

Munich Re. 2018. "NatCatSERVICE: Loss events worldwide 2017. Geographical overview." Accessed March 6, 2018. https://www.munichre.com/site/corporate/get/params_E976667823_Dattachment/1627370/MunichRe-NatCat-2017-World-Map.pdf.

NASEM (National Academies of Sciences, Engineering, and Medicine). 2018a. *Resilience in transportation planning, engineering, management, policy, and administration.* Washington, DC: National Academies Press.

NASEM. 2018b. *Renewing the national commitment to the interstate highway system: A foundation for the future.* Washington, DC: National Academies Press.

NOAA (National Oceanic and Atmospheric Administration). 2013. "National coastal population report: Population trends from 1970 to 2020." Accessed October 17, 2018. https://aamboceanservice.blob.core.windows.net/oceanservice-prod/facts/coastal-population-report.pdf.

OED (Oxford English Dictionary) Online. 2014. *Resilience.* Oxford: Oxford University Press.

Otwell, K., and B. Aldridge. 1989. "The role of vulnerability in risk management." In *Proc., 5th Annual Computer Security Applications Conf.,* 32–38. New York: IEEE Computer Society Press.

Porter, K., A. Wein, C. Alpers, A. Baez, P. Barnard, J. Carter, et al. 2011. *Overview of the ARkStorm scenario.* Open-File Rep. 2010-1312. Reston, VA: US Geological Survey.

Rosenzweig, C., W. D. Solecki, S. A. Hammer, and S. Mehrotra. 2011. *Climate change and cities: First assessment report of the urban climate change research network.* Cambridge, UK: Cambridge University Press.

Schröer, K., and M. R. Tye. 2019. "Quantifying damage contributions from convective and stratiform weather types: How well do precipitation and discharge data indicate the risk?" *J. Flood Risk Manage.* 12 (4): e12491.

Seneviratne, S. I., et al. 2012. "Changes in climate extremes and their impacts on the natural physical environment." In Chap. 3 in *Managing the risks of extreme events and disasters to advance climate change adaptation,* edited by C. B. Field, V. Barros, T. F. Stocker, D. Qin, D. J. Dokken, K. L. Ebi, et al., 109–230. Cambridge, UK: Intergovernmental Panel on Climate Change.

Stephenson, D. B. 2008. "Definition, diagnosis, and origin of extreme weather and climate events." In Chap. 1 in *Climate extremes and society,* edited by H. F. Diaz and R. J. Murnane. Cambridge, UK: Cambridge University Press, 11–23.

Sudhalkar, A., C. Chan, C. Bonham-Carter, and M. Smith. 2017. "C40 infrastructure and interdependencies and climate risks." Accessed October 17, 2018. https://unfccc.int/sites/default/files/report_c40_interdependencies_pdf.

Sweet, W. V., and J. Park. 2014. "From the extreme to the mean: Acceleration and tipping points of coastal inundation from sea level rise." *Earth's Future* 2 (12): 579–600.

Swiss Re. 2014. "Natural catastrophes and man-made disasters in 2013: Large losses from floods and hail; Haiyan hits the Philippines." Accessed March 27, 2014. http://media.swissre.com/documents/sigma1_2014_en.pdf.

Tredgold, T. 1828. "Article CS102127326, used in the Royal Charter of the Institution of Civil Engineers." *The Times,* London, June 30, 1828.

Tye, M. R., G. J. Holland, and J. M. Done. 2014. "Rethinking failure: Time for closer engineer–scientist collaborations on design." *Proc. Inst. Civ. Eng. Forensic Eng.* 168 (2): 49–57.

USGCRP (US Global Change Research Program). 2018. Vol. 2 of *Impacts, risks, and adaptation in the United States: Fourth national climate assessment,* edited by D. R. Reidmiller, C. W. Avery, D. R. Easterling, K. E. Kunkel, K. L. M. Lewis, T. K. Maycock, et al. Washington, DC: USGCRP.

Wahl, T., S. Jain, J. Bender, S. D. Meyers, and M. E. Luther. 2015. "Increasing risk of compound flooding from storm surge and rainfall for major US cities." *Nat. Clim. Change* 5 (12): 1093–1097.

Walker, B., C. S. Holling, S. R. Carpenter, and A. P. Kinzig. 2004. "Resilience, adaptability and transformability in social–ecological systems." *Ecol. Soc.* 9 (2): 5.

Wing, I. S., A. Z. Rose, and A. M. Wein. 2016. "Economic consequence analysis of the ARkStorm scenario." *Nat. Hazard. Rev.* 17 (4): A4015002.

Yohe, G., and R. Leichenko. 2010. "Chapter 2: Adopting a risk-based approach." *Ann. N.Y. Acad. Sci.* 1196 (1): 29–40.

CHAPTER 2

State of the Science in Meteorological/Hydrological Extremes

2.1 INTRODUCTION

The first key message from the Fourth National Climate Assessment (NCA4) report of the US Global Change Research Program (USGCRP) (Hayhoe et al. 2018) is that the rate of global climate change substantially outpaces any climatic effects because of natural variability. These changes are manifested in a variety of ways, but the most visible and verifiable indicator is an increase of 1.8 °F in mean global temperatures during the period from 1901 to 2016 that cannot be attributed to natural phenomena. The primary cause of global climate change stems from the emission of greenhouse gases (GHGs). Hayhoe et al. (2018) find that even if significant reductions in GHGs do occur, global temperatures are projected to increase by as much as 3.6 °F beyond the year 2050 compared with preindustrial temperatures; without such reductions, this number jumps to 9.0 °F or more by the year 2100.

   In addition to the direct impacts of higher temperatures on critical infrastructure (e.g., increased energy demand for cooling purposes, damage to roads and rail, degradation in water quality), such extreme upward movements in temperature can affect other meteorological and hydrological factors that, in turn, can have substantial impacts on civil engineering infrastructure. Such factors include precipitation, streamflow and runoff, frequency and intensity of droughts and fires, storm tracks, frequency and intensity of tropical cyclones (TCs), rise in sea levels and surface temperatures, storm surges, ocean acidity, sea ice extent, frequency of atmospheric icing events, ocean wave height and period, snowpack duration and depth, and the frequency and cost of natural disasters, among others. The projected influence of global warming on many of these factors will be discussed in the following sections.

2.2 ATTRIBUTION OF EXTREME EVENTS TO CLIMATE CHANGE

The attribution of specific meteorological and hydrological extreme events to global warming is challenging, although efforts are made to do so (Herring et al. 2019). Changing climate imparts not only a new *normal* but also new probability thresholds of extreme events. Although change in mean climate trends can be readily associated with anthropogenic drivers of the climate system, tying the extremes to increased concentrations of GHGs has been more elusive because of the complex meteorological interactions that, when combined, create severe weather events such as hurricanes, floods, hail, and blizzards. The need to use observational records of sufficient duration to capture the most extreme events and estimate their rarity (or return frequency) creates two complications. First, not all locations have observations of sufficient longevity and reliability to accurately estimate event frequencies. Second, these estimates are premised on statistical stationarity, whereby the statistical properties are assumed to be constant over the time series, which no longer can be assumed to be accurate (Milly et al. 2008). Although nonstationarity in the historical precipitation record is challenging to establish (Sun et al. 2018), historical temperatures show a clearer stochastic trend that is tied to radiative drivers such as CO_2 (Kaufmann et al. 2010). Similarly, compound extremes such as drought, which is dependent on both temperature and precipitation, have demonstrated considerable variability in the last 1,000 years, which has complicated the detection of trends and attribution of specific events (Coats et al. 2015; Wehner et al. 2017). Assumptions of stationarity become less tenuous further into the future. One method of minimizing the influence of nonstationarity on future assessments of climate change impacts is to use progressive snapshots in time (e.g., 10 to 30 year periods). However, the statistical uncertainty of low-frequency extremes increases greatly with smaller time samples, especially where the return periods are far beyond the actual sampling period length.

Large-scale and long-term weather patterns, that is, synoptic climatology, which are associated more with climate change than with extreme events, also impact local weather responses. Therefore, assessing the impact of climate change on these patterns is important. For example, large-scale upper-level flows in the atmosphere can block storms from entering certain regions and enhance droughts or heat waves (Sillmann and Croci-Maspoli 2009) while creating favorable meteorological environments for severe storms to form elsewhere (Prein et al. 2016). In the case of Hurricane Harvey, analyses of the prestorm atmospheric water environment surrounding the storm were shown to be unlikely without increased long-term atmospheric and ocean water temperatures (Van Oldenborgh et al. 2017, Trenberth et al. 2018). Statistical analyses of a growing global weather record provide insight into changes in the likelihood of extreme events by showing an increased probability of local events.

With advances in meteorological analysis and computational ability, climate scientists can also better assess how anthropogenic climate forcing impacts the frequency and intensity of disruptive events through multiple computer simulations (NASEM 2016, Stone et al. 2013). Because of increasing computing speed and power, numerical simulations of weather and climate permit more detailed simulations with higher spatial resolution combined with improved techniques to *downscale* larger global simulations to regional scales. This provides the ability to compare *counterfactual* (i.e., omitting anthropogenic climate and land-use changes) versus existing or future climate change scenarios for the same evaluation period. The increase in computational power also allows the generation of multiple simulations (or *ensembles*) of a given event or climate scenario to facilitate multiple probabilistic analyses of a given scenario and provide an improved spectrum of the likelihood of events.

2.3 TEMPERATURE

The most visible and verifiable indicator of climate change is the increase in mean global temperatures stemming from the emissions of GHGs (Hayhoe et al. 2018). Effects of increasing temperatures have resulted in an increase in the length of the frost-free season in all regions of the United States since the early 1900s, a reduction in the frequency of cold waves since the early 1900s, and an increase in the frequency of heatwaves since the mid-1960s (Hibbard et al. 2017, Vose et al. 2017).

Vose et al. (2017) find that projections of annual mean temperature within the United States range between increases of 2.3 and 11.0 °F by the late-twenty-first century relative to 1986 to 2015 depending on the emission scenarios employed (i.e., RCP4.5 versus RCP8.5; see Figure 2-1). In terms of more extreme temperatures, the average number of days below freezing is projected to decrease, whereas the number of days above 90 °F will likely increase.

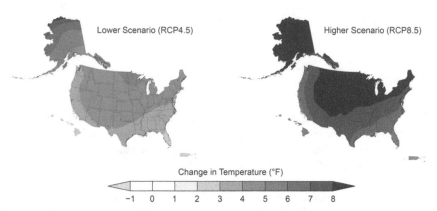

Figure 2-1. Projected changes in mean annual temperature in the United States by the late-twenty-first century relative to the period 1986 to 2015.

Source: Hayhoe et al. (2018); reproduced from CICS-NC and NOAA NCEI.

In addition to the intensity of temperature extremes, their persistence must also be considered. Heat waves are broadly described as extended periods of high temperatures that have significant impacts on both the natural and the built environments (Perkins 2015). In addition to detrimental impacts on ecosystems and increasing energy consumption, heat waves can cause desiccation and cracks in earthen infrastructure, as discussed in Chapter 3. In last few decades, the frequency and intensity of heat waves have increased substantially throughout the United States (Keellings and Moradkhani 2020) and across the globe (Perkins 2015, Raei et al. 2018). Heat waves are defined as one of two types: relatively dry daytime heat waves (Type I) or humid nighttime events (Type II) that are associated with higher human morbidity and mortality (Gershunov and Guirguis 2012). Both types of heat waves are projected to increase in intensity over the coming decades, although Type II events are projected to increase at a faster rate, of which there is already evidence in several subregions of California (Gershunov and Guirguis 2012). Vose et al. (2017) also project that heatwaves will become more intense, whereas cold waves will become less intense in the future.

Although climate change impacts are commonly associated with increasing high-temperature extremes as previously described, and despite the overall decreasing frequency of cold extremes, there have also been periods of enhanced cold extremes. Such instances often coincide with major volcanic eruptions and various mechanisms of natural climate variability, such as major El Niño/Southern Oscillation (ENSO) episodes (e.g., extreme El Niño of 1997 to 1998; Johnson et al. 2018). More recently, the *temperature hiatus* (2002 to 2014) recorded an increased frequency of cold extremes without the occurrence of a major volcanic eruption or a strong ENSO event. Johnson et al. (2018) found that these recent cold extremes are closely linked to positive extremes in the midtropospheric (500 hPa/7.25 psi) geopotential height anomalies over the high latitudes and negative anomalies over the North Atlantic and Eurasia. The geopotential anomalies are also related to a *warm Arctic-cold continents* pattern, which has been linked to recent trends in midlatitude extreme wintertime cold temperatures (Kug et al. 2015, Sun et al. 2016).

2.4 PRECIPITATION

Rising global temperatures in response to anthropogenic influences have contributed to a corresponding increase in the intensity and frequency of extreme precipitation owing to the atmosphere's increasing capacity to hold water (O'Gorman 2015, Prein et al. 2017, Westra et al. 2014). Average annual precipitation has increased by about 4% throughout the United States since 1901, with regional changes ranging from an increase of more than 30% in Alaska to a decrease of 20% in the Southwest (Easterling et al. 2017, Prein et al. 2016).

Higher temperatures lead to higher evaporation rates on the ground, which increase the overall water vapor content of the atmosphere. The result is an increase in the frequency and intensity of heavy precipitation events. Consequently, it has been found that the frequency and intensity of heavy precipitation have increased more than average precipitation (Easterling et al. 2017). For example, the observed extreme events characterized by daily precipitation above the 99th percentile have increased by 55% and 42% since 1958 in the Northeast and Midwest, respectively (top-right plot in Figure 2-2) (Easterling et al. 2017), and are projected to increase by an additional 40% for the period 2070 to 2099 when compared with 1986 to 2015. Other metrics given in Figure 2-2 include historic trends in 5 year maximum daily precipitation since 1901 (top-left plot) and in the number of 5 year, 2 day events since 1901 (bottom-left plot) and since 1958 (bottom-right plot). All metrics show maximum increases in the Northeast and Midwest and little change or decreases in the Southwest.

Other studies have found similar results of little or no change in the frequency of extreme precipitation events in the western United States over the last several decades (Bonnin et al. 2011, Lehmann et al. 2015). In contrast to these

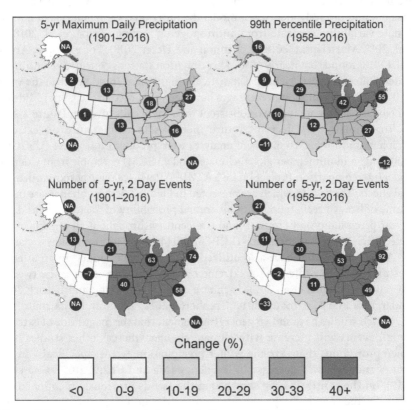

Figure 2-2. Percent observed changes in the magnitude of 5 year maximum daily precipitation (top-left) and the number of 5 year, 2 day events (lower-left) for the period 1901 to 2016 and 99th percentile precipitation (top-right) and number of 5 year, 2day events (lower-right) for the period 1958 to 2016 throughout the United States.

Source: Easterling et al. (2017); reproduced from CICS-NC and NOAA NCEI.

findings, trends in more extreme events have been found to be positive in the West. For example, increasing trends in daily, 5% annual probability (20 year), seasonal precipitation have been found (not given) regardless of season, which demonstrates the importance of the specific definition that is used to define an extreme event when considering infrastructure vulnerability.

The exact trajectory of historical trends in short- and long-term meteorological extremes observed in many locations throughout the world has also been found to rely heavily on the natural variability of the climate system (Cheng et al. 2018, Giovannettone and Zhang 2019, Kendon et al. 2018, Liu et al. 2018, Schubert et al. 2016). Several studies have looked at the impacts of natural climate variability through hydroclimate indexes such as the ENSO (Curtis and Adler 2000, Trenberth and Stepaniak 2001), the Madden–Julian Oscillation (Madden and Julian 1994), the North Atlantic Oscillation (Barnston and Livezey 1987, Hurrell 1995), the Pacific Decadal Oscillation (PDO) (Zhang et al. 2017), and the Arctic Oscillation (Vavrus et al. 2017), among others. A major objective in the field is to tease out and properly identify or model the individual contributions of natural climate variability separate from anthropogenic forcing (Armal et al. 2018, Li et al. 2017, Martel et al. 2018, McKinnon and Deser 2018). For example, Armal et al. (2018) found that nearly 60% of precipitation stations considered within the United States have experienced a statistically significant trend between the years 1900 and 2014; the highest percentages have been observed in the Upper Midwest and Northeast climate regions, consistent with the results given in Figure 2-2.

To estimate the effects of future increases in temperature on extreme precipitation, future extreme event analyses were produced for NCA4 (USGCRP 2018) using a multimember global climate model (GCM) ensemble from Coupled Model Intercomparison Project Phase 5 (CMIP5) downscaled to accommodate for local effects. The results, which are presented in Figure 2-3, show an increase in the volume of heavy precipitation with an annual probability of 5% (defined as daily, 20 year precipitation) out to mid- and late-century throughout all regions of the United States using both moderate- (RCP4.5) and high- (RCP8.5) emission scenarios.

Additional studies have also utilized various climate models in an attempt to estimate how the intensity of extreme rainfall will increase over the twenty-first century (Collins et al. 2013, Farnham et al. 2018, Huang and Ullrich 2017, Pendergrass and Hartmann 2014, Trenberth 2011, Zhou and Khairoutdinov 2017). Such studies have shown general agreement that the magnitude of extreme rainfall events will increase with climate change, whereas other studies have shown projections that extreme winter precipitation events, specifically in the form of snowfall, will decrease in frequency (Ma and Chang 2017). Gao et al. (2017), on the contrary, have shown that the trends previously mentioned are regionally and seasonally specific; for example, the authors project the frequency of heavy winter precipitation to become more frequent along the Pacific Coast of California, whereas the frequency of heavy summer precipitation is projected to decrease in the Midwest.

Downscaled CMIP5 analyses also show that overall rainfall events will be distributed to slightly more frequent dry days punctuated by more frequent

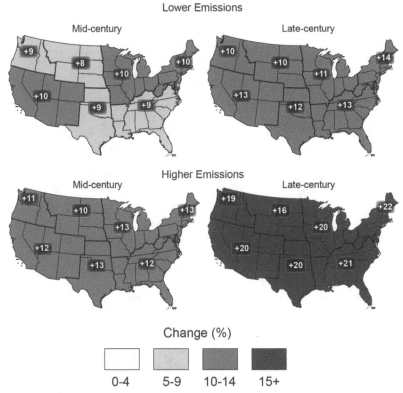

Projected Change
in Daily, 20-year Extreme Precipitation

Figure 2-3. Projected change in 20 year return daily rainfall magnitudes for the continental United States by region. Shown are mid- to late-twentieth-century periods for both moderate- (RCP4.5) and high- (RCP8.5) emission scenarios used in CMIP5. The analysis was performed using the LOCA downscaling method.

Source: Easterling et al. (2017); reproduced from CICS-NC and NOAA NCEI.

heavy rainfall events, with moderate spectrum events decreasing in frequency (Wehner et al. 2017); this is especially true on the West Coast of the United States (Dettinger 2016, Gershunov et al. 2019). Examples provided by Gershunov et al. (2019) reveal projected patterns for three river basins, along the northern, central, and southern West Coast of an increased frequency of dry days and decreased frequency of medium-intensity (30th to 80th percentile) events under RCP8.5 for the latter half of the twenty-first century (Figure 2-4). Decreases in lower-intensity events have also been projected for two of the basins [Figures 2-4(a and b)], whereas the frequency of extreme events (>90th percentile) has been projected to increase substantially in all river basins. This increase is explained by the projected increase in the frequency of atmospheric river (AR) events, which

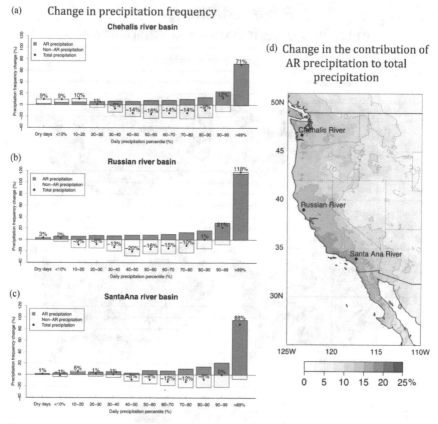

Figure 2-4. Projected changes in the frequency of all percentiles of daily precipitation for the (a) Chehalis, (b) Russian, (c) Santa Ana River basins; projected changes in the contributions of precipitation events linked (shaded) and not linked (unshaded) to AR activity are also shown. (d) The spatial distribution of projected changes in the contribution of AR precipitation to total precipitation is shown for the western United States.

Source: Gershunov et al. (2019); permission to reproduce under the Creative Commons License: https://creativecommons.org/licenses/by/4.0/legalcode.

present the major contribution to extreme precipitation along the West Coast, along the entire coast [Figure 2-4(d)].

In contrast, the frequency of non-AR events, which predominantly include low- and medium-intensity events, is projected to decrease because of increasingly anticyclonic circulation over the northeastern Pacific, a northward shift in storm tracks, and a lower frequency of impacts by midlatitude cyclones. The resulting volatility of such precipitation patterns (increased precipitation resulting from fewer precipitation events) will challenge water resources and risk management responsible for balancing the mandates for water storage and flood control.

An increase in flooding corresponding to the projected increases in the intensity and frequency of extreme precipitation events is more difficult to predict, even though extreme precipitation is a major cause (Hayhoe et al. 2018). The frequency and intensity of flood events depend on several other compounding factors, including land use and water resource management operations. Therefore, anthropogenic warming has not been conclusively identified as a major indicator of the projected changes in flood events (Wehner et al. 2017).

The coarse-resolution climate models underestimate the projected increase in the magnitude of extreme precipitation (Allan and Soden 2008, Chen et al. 2018, Li et al. 2018). It has been shown that increasing the atmospheric resolution through downscaling improves their ability to simulate the spatial extent, intensity, and seasonal timing of such events (van der Wiel et al. 2016). Typically two types of downscaling techniques exist: (1) statistical downscaling based on relationships between large-scale atmospheric variables and regional-scale observations or reanalysis data (Sharma et al. 2019) and (2) dynamical downscaling that utilizes higher-resolution regional climate models (RCMs) to extrapolate the effects of large-scale climate processes to the scale of interest. The relationships developed in statistical downscaling are obtained using one of a variety of methods, including linear and partial least-squares regression, artificial neural networks, and other machine learning techniques, relevance vector machines, and so on. Regardless of the specific method used, the primary advantage of statistical downscaling is that it requires less computational energy and time to implement.

A major concern implicit in the use of statistical downscaling when considering the potential impacts of climate change is the use of relationships that are tuned to the current climate and that may not remain constant over time. Alternatively, dynamical downscaling presents a more accurate method by which to model changes in extreme events through a physically based approach (Komurcu et al. 2018). Dynamical downscaling allows a high-resolution representation of topography and coastlines and can resolve smaller-scale processes such as convective precipitation, which allows an improved prediction of extreme precipitation and flooding events. Convective-permitting models (CPMs), for example, have been found to be able to estimate subdaily extreme precipitation intensity and extent with more skill in contrast to lower-resolution parameterized convection models (Chan et al. 2018, Prein et al. 2015, Wang et al. 2018, Zhou and Khairoutdinov 2017). The disadvantage of using CPMs, as with many other types of dynamical downscaling techniques, is that they are computationally intensive, and therefore, it is difficult to estimate uncertainties associated with the model other than those arising from resolution. For instance, the computational cost prohibits running multiple realizations from different GCM models to estimate uncertainty in the initial conditions. In addition, the boundary conditions used, domain size, and location can have profound influences on the downscaled model output and biases (Done et al. 2015, Prein et al. 2013). Prein et al. (2017) present one example of a CPM simulating hourly precipitation extremes to demonstrate that the relationship between increasing temperature and precipitation intensity varies regionally and is dependent on moisture availability. Although the consensus is that the frequency and intensity of extreme precipitation will increase throughout the United States and the twenty-first century (Allan and Soden 2008, Min et al. 2011), it is also apparent that the nature of the changes will vary regionally and seasonally driven by different weather systems (Prein et al. 2016).

The projected changes in precipitation are as difficult to discern from natural and model variability in many regions as the trends in historical precipitation. Figure 2-5 shows that the only regions and seasons where the projected changes are distinguishable from the range of natural variability (indicated by red dots) are the Upper Midwest, the Northeast, and the State of Alaska, during winter and spring (Hayhoe et al. 2018). Areas that are hatched (South and West) are characterized by small and insignificant changes, whereas all other areas exhibit projected changes that, although significant, are much smaller than the range of natural variability. Projected changes during summer and fall (not given) are insignificant and much less than the amplitude of natural variability throughout all regions of the country. Such comparisons between projected impacts of climate change and the range of natural climate variability are important because of that most critical infrastructure should already be designed to withstand the effects of natural interannual shifts. It is recommended that adaptation efforts to accommodate precipitation extremes focus on the Upper Midwest and Northeast where projected trends indicate that future precipitation behavior may exceed past experience.

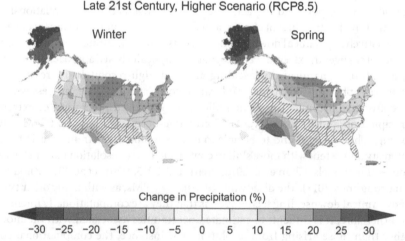

Late 21st Century, Higher Scenario (RCP8.5)

Winter

Spring

Change in Precipitation (%)

-30 -25 -20 -15 -10 -5 0 5 10 15 20 25 30

Figure 2-5. Projected changes in precipitation throughout the United States for the period 2070 to 2099 compared with the period 1986 to 2015 using the RCP8.5 scenario for the winter and spring seasons. The areas with red dots indicate that projected trends are a large fraction of natural variability, whereas hatched areas indicate trends that are both small and insignificant. All other areas are characterized by significant trends that are a small fraction of natural variability.

Source: NCA4 (Hayhoe et al. 2018).

2.5 SNOWPACK AND SNOWMELT

Warmer temperatures have been linked to changes in snowpack thickness and density [snow-water equivalent (SWE)] and snowmelt timing (Wehner et al. 2017). Observations in many areas have already demonstrated that increased temperatures are resulting in thinner and denser snowpack owing to an increase in the percentage of wintertime precipitation in the form of rain rather than snow and earlier accumulated snowmelt (Stewart 2009). Future projections indicate a continuation of this behavior. For example, Mastin et al. (2011) showed that projections of temperature and precipitation over eight different snowmelt-dominated basins in the West and Northeast resulted in decreases in the SWE of snowpack ranging from approximately 12% to 81% between 2006 and 2090. The decrease in SWE is expected to vary spatially with respect to elevation; the greatest reductions will occur at lower elevations where critical temperature thresholds are more easily exceeded (MacDonald et al. 2011, Mastin et al. 2011). With increased urban development already exacerbating the challenges associated with water allocation particularly during periods of water shortage (e.g., San Francisco, California, in 2006 to 2007), the magnitude of the described projections suggests that the effects of climate change will only add to the burden of balancing competing water demands.

As most of the wintertime precipitation is stored as snow in the upper elevations of snowmelt-dominated (in contrast to rain-dominated) basins, a large percentage of the annual runoff occurs during spring and early summer as a *runoff pulse* caused by the release of the snowpack. Changes in the SWE during the winter and spring months, whether owing to high interannual variability in temperature and precipitation or anthropogenic warming during the winter and spring months, can have a substantial impact on the timing and amount of snowmelt runoff contained within this pulse. In fact, it has been found that the runoff pulse can contribute up to 75% of annual runoff in snowmelt-dominated basins of the western United States (Dettinger 2005, Stewart et al. 2004). In particular, peak runoff has been observed to occur increasingly earlier (Dettinger 2005, Fritze et al. 2011, Stewart et al. 2004) because of thinner snowpack (Mote et al. 2005) and higher percentages of precipitation falling as rain rather than snow (Knowles et al. 2006). Figure 2-6 shows that the onset of spring snowmelt (measured in terms of the date of initial surge to streamflow) [Figure 2-6(a)], as well as the timing of the center of volume of annual streamflow [Figure 2-6(b)], had shifted by as much as 3 weeks earlier by the beginning of the twenty-first century compared with the mid-twentieth century; an even greater shift of up to 35 days earlier has been projected for the late-twenty-first century (Stewart et al. 2004). Such variability creates uncertainty in terms of water supply, which, in turn, makes water allocation for irrigation, municipal demands, inland navigation, hydroelectric power, and environmental protection (e.g., salmon migration) more challenging (Barnett et al. 2005). This is discussed further in Chapter 3.

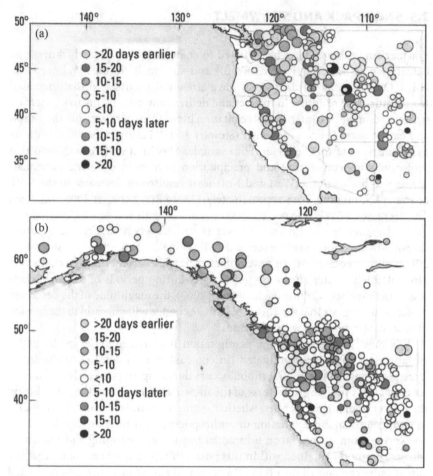

Figure 2-6. Changes (a) in the timing (days) of the onset of spring snowmelt in the western United States, and (b) in the date of the center of volume of annual streamflow for the beginning of the twenty-first century compared with the mid-twentieth century.

Source: USGS (2005).

2.6 DROUGHT

Drought can occur across a multitude of timescales and can have severe impacts on natural ecosystems and landscapes, surface and groundwater supplies, and agricultural production, and can result in major socioeconomic disruptions. Droughts are typically classified as meteorological (precipitation deficit), agricultural (deficit in soil moisture and low plant health), hydrological (deficit in surface water, groundwater, and/or reservoir levels), socioeconomic (impact on supply and demand of goods and services), snow drought (because of low precipitation and/or low snowpack), or broadly anthropogenic drought. The latter corresponds to drought events caused or intensified by human activities such as resource/infrastructure mismanagement (AghaKouchak et al. 2015). In addition to the preceding classifications, any type of drought that develops much more rapidly than normal is also referred to as a flash drought (PaiMazumder and Done 2016, Otkin et al. 2018). The onset of flash drought is often the result of the simultaneous occurrence of multiple hazards (e.g., low precipitation, heat waves, high atmospheric evaporative demand) that can result in either of the drought types previously listed. A combination of high temperatures, sunny skies, and below-normal precipitation was responsible for the sudden onset of the widespread drought that spread across the central United States in 2012 (Fuchs et al. 2015). Many locations went from near-normal to extreme drought conditions in less than 2 months. Agricultural losses owing to this event exceeded $30 billion across the United States (NOAA NCEI 2017). For example, the California–Nevada drought of 2012 to 2015 was a result of a combination of low precipitation and high evaporative demand caused by higher-than-normal average temperatures. In addition to the effects on agriculture, flash droughts can cause substantial increases in fire risk above that of a typical drought over an impacted area.

Projected trends in temperature are expected to have a significant impact on evaporation, which will severely impact the water budget, imposing long-term deficits of water (Collins et al. 2013, Dai 2013, Wehner et al. 2017). As a result, surface soil moisture availability, which is a critical diagnostic metric for drought, is projected to decrease (Hayhoe et al. 2018) and lead to an increase in drought caused by evaporative demands. Projections of near-surface moisture for the late-twenty-first century using the RCP8.5 emissions scenario are given in Figure 2-7. Near-surface soil moisture is projected to decrease throughout the entire United States, which implies that future droughts are projected to become more frequent and intense, and persist for longer periods. In combination with lower soil moisture, decreased precipitation in the West and Southwest will further exacerbate drought frequency, intensity, and duration. Hatchett et al. (2015) demonstrated that the low precipitation anomalies experienced during the 2012 to 2015 California–Nevada drought were within the range of natural climate variability, but when temperature anomalies (and hence evaporative demand) were included, the average severity of the drought was more extreme than the severity of past extended periods of drought. In contrast, Sheffield et al. (2012) found little change in drought frequency and intensity at a global scale over the last several decades when they accounted for the physical parameters excluded by the Palmer Drought Severity Index (e.g., humidity, available energy, and wind speed). This emphasizes that there is much uncertainty surrounding past trends in drought frequency and intensity that depend on the metric used to measure drought. Other glaring issues that must be addressed when assessing the historic and potential future impacts of climate change on drought are the availability of more comprehensive and consistent precipitation data sets and a

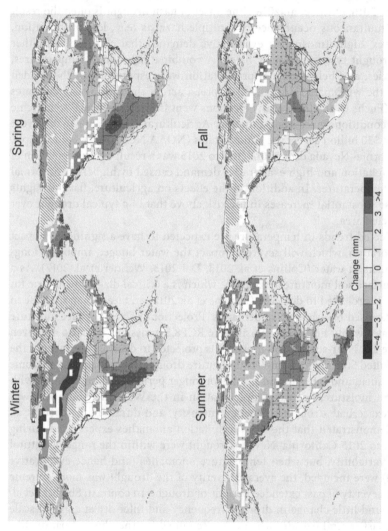

Figure 2-7. Projected change in large-scale near-surface soil moisture over North America for the late-twenty-first century using the CMIP5 high emission scenario (RCP8.5).

Source: CISS NOAA-National Centers for Environmental Information (NCEI) and Cooperative Institute for Climate and Satellites-North Carolina (CICS-NC).

better understanding of the role of the various dominant mechanisms of climate variability, such as ENSO and the PDO, in precipitation patterns (Giovannettone 2021, Trenberth et al. 2014).

There is overall wide consensus in the climate community that the frequency and intensity of droughts, including flash droughts, will increase with climate change throughout the United States (Bertrand and McPherson 2018, Spinoni et al. 2018, Vavrus et al. 2017). Regions of particular concern include the Southwest and Southeast (Herrera-Estrada and Sheffield 2017), the western United States (Wehner et al. 2011), and the Midwest (Mishra et al. 2010). However, there is high uncertainty in the magnitude of these changes (Herrera-Estrada and Sheffield 2017) owing to projected increases in temperature and precipitation, as well as the character of the precipitation (Mishra et al. 2010, Seager et al. 2009) during the winter months. Expected trends in drought occurrence have already been observed in many locations throughout the world (Crockett and Westerling 2018, Ma et al. 2017, Shao et al. 2018). For example, Crockett and Westerling (2018) noted that the core of droughts in the western United States has increased in size when compared with the last 100 years. On the contrary, Mo and Lettenmaier (2018) found that severe droughts that cover more than 50% of the continental United States are occurring less often, throughout the observational record, and are less severe as time progresses.

Water supply is particularly vulnerable to changes in winter snowfall (i.e., snow drought), especially in the Mountain West. Increasing importance is being placed on the need to differentiate dry snow droughts (i.e., a lack of wintertime precipitation) and warm snow droughts (i.e., higher-than-normal temperatures generating rain instead of snow) that also lead to earlier snowmelt (Harpold et al. 2017). Dry snow drought results in both low flows during wintertime and a lack of springtime pulses caused by snowmelt, whereas warm snow droughts are characterized by large snowmelt pulses during wintertime and low flow during springtime. Both present unique challenges for water managers in terms of accommodating the large snowmelt pulses to avoid flooding while also attempting to accommodate the following springtime low flows.

With increased drought frequency come associated impacts such as poor soil quality and increased erosion potential. Increased soil drying at the surface can lead to crusting and degradation of topsoil, which varies with soil texture. This, in turn, can change soil response to rainfall, that is, erosion. Given the expected increase in heavy rainfall frequency in coming decades, there is concern that sediment deposition into streams will increase, because a single heavy rainfall event can impart much of a season's sediment deposition. This has major implications for identifying potential impaired rivers and streams as well as costs of dredging lakes and reservoirs (Li and Fang 2016, Nearing et al. 2004, Segura et al. 2014).

In addition to the factors related to the frequency and intensity of drought previously discussed, the concurrent occurrence of droughts and heat waves needs to be considered. A recent study showed that droughts have been warming faster than the average climate across most of the United States (Chiang et al. 2018).

The relationship between rising temperatures and land–surface interactions can potentially intensify the severity and impacts of both future droughts and heat waves (Chiang et al. 2018). A study of the last five decades showed that concurrent droughts and heat waves increased substantially across most of the United States (Mazdiyasni and AghaKouchak 2015). Given the projected increases in temperatures, combined with the increased atmospheric moisture content (IPCC 2013, USGCRP 2018), more frequent concurrent drought–heatwave extremes are expected in the future, which can have significant impacts on the built environment and human health. It is, therefore, critical to improving our understanding of the potential impacts of droughts and compound droughts and heat waves on the built environment (e.g., levees and earthen dams; Vahedifard et al. 2016) and human responses (e.g., increased energy demand arising from air-conditioning use).

Drought and heat waves, especially when they occur together, can weaken earthen infrastructure such as embankments, levees, and dams. The potential weakening mechanisms include, but are not limited to, desiccation cracking, enhanced erosion, land subsidence, fissuring, reduced soil strength, and soil softening (ASCE CACC 2015, Brooks et al. 2012, Vahedifard et al. 2016). Current design codes and guidelines do not explicitly account for droughts and heatwaves or their future projections. However, given the expected changes in their frequency and severity, there is a critical need to develop methodological frameworks for integrating drought and heat wave information in engineering risk assessment and design concepts, both for operational purposes (e.g., passive building cooling) and for potential structure failure mechanisms.

2.7 WILDFIRE

Wildfire is an increasing threat to infrastructure in the western United States and Alaska (Wehner et al. 2017). Between 1987 and 2003, the frequency of western US wildfires larger than 400 ha in area increased fourfold compared with 1970 to 1986, and the area burned increased by a factor of 6.5 (Dennison et al. 2014, Westerling et al. 2006). These changes have been linked to the combined influences of a changing climate (Littell et al. 2009, Westerling et al. 2006), twentieth-century fire suppression policies (Wehner et al. 2017), and increased human settlement within and adjacent to forests (Martinuzzi et al. 2015). In the last decade, the risk of large fires has likely increased by 33% to 50% in Alaska (Partain et al. 2016), and the season may be lengthening (Flannigan et al. 2009). In addition to the increased risks to life safety, health, and property damage from wildfire, the cost of wildfire suppression in the western United States has soared in recent years (Vose et al. 2018). In 1995, wildfire-related expenses consumed only 16% of the total US Forest Service budget (approximately $2.4 billion), whereas these expenses consumed 52% of the total budget (approximately $4.8 billion) for the fiscal year 2015 (USDA 2015).

The contribution of the climate to wildfire ignition, intensity, and spread is dictated by precipitation, temperature, and wind patterns, which, in turn, affect vegetation type/distribution (Ryan 2000, Williams et al. 2010). The impact of future climate change (as manifested through changes in each of these variables) is expected to vary greatly across the United States. In the northern Rockies, wildfire frequency is projected to increase under higher warming scenarios, such that years without large fires may become rare by the mid-century (Westerling et al. 2011). Very large fires that exceed 50,000 acres may increase across the western United States by the mid century under both higher and lower emission scenarios (Stavros et al. 2014). In Alaska, the risk of severe wildfires is projected to increase by up to a factor of 4 by the end of the century even under moderate emission scenarios (Young et al. 2017), the season is likely to continue to lengthen, and the area burned may increase between 25% and 53% (Joly et al. 2012).

Projected changes in lightning frequency will also affect wildfire ignition/ frequency. Romps et al. (2014) predicted the potential for an increase in lightning over the continental United States of approximately 12% through the twenty-first century because of increased convective potential. Comparing cloud electrification parameterizations, Finney et al. (2018) showed considerable regional variability with small increases over much of the continental United States (including Alaska) and high uncertainty over the central United States. In the Southeastern United States, Prestemon et al. (2016) projected a 30% increase in the annual area burned by lightning-ignited wildfires by 2060. Because this is a region where development is increasing in and adjacent to forested areas, these changes will likely increase risks to life safety, infrastructure, and property and degrade regional air quality (Vose et al. 2018).

Projected changes are also expected to impact the length and seasonality of the wildland fire season with many regions in the United States experiencing longer fire seasons (Liu et al. 2013, McKenzie and Littell 2017, Riley and Loehman 2016, Terando et al. 2016, Westerling 2016). Indeed, some areas of the United States are now experiencing significant wildfire activity all-year-round. For example, the 54,023 acre Legion Lake Wildfire occurred in December 2017 in the Black Hills of South Dakota. The peak fire season in this region of the United States typically runs from June to August, making a large December wildfire incredibly rare (Clabo 2018). GCM projections also suggest that a shift in seasonality will occur in other regions. The peak fire season, for instance, historically occurs in autumn in southern California after the onset of the Santa Ana winds, which tend to exacerbate already dry conditions and spread whatever fires may ignite. Guzman-Morales and Gershunov (2019), however, projected that Santa Ana winds affecting the region will reduce in frequency, particularly in the autumn and spring periods, thus shifting the peak fire season to the winter months.

Secondary effects are observed from climate change that also affect wildfire activity. A reduction in cold-temperature days creates a more favorable environment for endemic yet destructive insects such as the Mountain Pine Beetle

(MPB) (*Dendroctonus ponderosae*). Widespread tree mortality across the western United States has been attributed to the MPB and other bark beetles (Bentz et al. 2010). It has been shown that tree mortality owing to these disturbances affects the fire regime of particular ecosystems (Simard et al. 2011) and may counteract the direct effects of climate change to some effect through a reduction in fine fuel growth (Littell et al. 2010). Recent research has also demonstrated that fire behavior differs with the different stages of MPB infestation (Jenkins et al. 2014). During the early stages of infestation, increased flammable terpenoid production increases the risk of eruptive fire behavior (Jolly et al. 2012, Page et al. 2012). During the later stages of infestation, larger stands of dead, woody fuels reduce the efficacy of fire suppression activities and the development of safe access and egress (Jenkins et al. 2014).

With continued human population increases, there is concern over the encroachment of the built environment into natural environs or an expansion of the wildland–urban interface (WUI). Within the continental United States, the WUI covers nearly 10% of the land area but contains approximately 38% of the housing units; California alone has over 5 million homes in the WUI (Radeloff et al. 2005). From 1990 to 2010, the number of homes and people within the continental US WUI grew by 12.7 and 25 million, respectively (Radeloff et al. 2018). Nevertheless, the approximate doubling of wildfire occurrences in the western United States between 1984 and 2015 can be attributed only to increased fuel aridity from anthropogenic climate forcings and not to human activity specifically (Abatzoglou and Williams 2016). The effect of enhanced human activity along the WUI was, rather, an increase in their exposure to the impacts of wildfire.

Not only do projected trends in wildfire frequency, extent, and intensity pose direct risks to human lives and the built environment, but also indirect risks exist through the lasting effects of wildfire activity on the landscape. The hydrology of burned watersheds dramatically increases the rainfall–runoff ratio, resulting in higher flood flows [on top of increasing projections owing to trends in extreme precipitation (Section 2.4)] (Moody and Martin 2001), sediment load, and debris flows (Cannon 2001) that may significantly increase flood risk to downstream infrastructure and private property for years to decades postfire. In a warming climate, postfire watersheds may no longer support germination and the establishment of historically present vegetation species, resulting in long-term changes to species and/or community structure (McKenzie et al. 2004). The replacement community may have different hydrologic properties and wildfire regimes. In permafrost regions, wildfires that significantly diminish the thickness of the soil organic layer result in postfire landscapes with thicker permafrost active layers and the formation of zones of continuously thawed soil (taliks) between the active layer and the underlying permafrost (Yoshikawa et al. 2003). Permafrost thaw from this and other mechanisms is likely to significantly increase damage to Alaska's infrastructure (Melvin et al. 2017).

2.8 STORM TRACKS AND MIDDLE-LATITUDE CYCLONES

Cold season extreme events are often delivered through large-scale middle-latitude cyclones. These events include coastal storm flooding that can be of the same magnitude or greater than tropical systems, icing events, blizzards, and marine windstorms. Changes to these storm systems also impact snowpack and other cold season precipitation, which can be important to the hydrologic cycle.

Although the surface lows that characterize these systems can be small compared with general circulation models' grid scales and resolutions, their intensity, points of origin, and storm paths are influenced by larger upper air patterns in the middle and upper troposphere. In combination with topography, sea temperatures, and land-sea features, these systems often form and travel along well-recognized paths or *storm tracks*. North American examples of these tracks include "Alberta Clippers," "Colorado Lows," and "Nor'easters."

Large-scale middle-latitude cyclone tracks have had an observable shift northward in recent decades, which is attributed to a poleward expansion of the tropics into middle-latitude regions (Bender et al. 2012; Wang et al. 2006, 2013). Given recently observed changes in large-scale weather patterns, it has been posited that climate change over the Arctic triggers amplified atmospheric waves that drive midlatitude storm systems and associated extreme weather (Francis and Vavrus 2012, Screen and Simmonds 2014, Semmler et al. 2018). Higher-wave amplitudes are indicators of atmospheric blocking patterns that drive extended warm and cold spells. Although future climate projections differ in the specific placement of storm tracks and their intensity, modeled pressure variations imply a reduced overall intensity of midlatitude cyclones and a northward trend in both the tracks and their origin point (Chang 2013, Colle et al. 2013).

2.9 TORNADOES AND HAIL

Attribution of tornado- and hail-generating storms and other more localized events to climate change and assessment of likely changes is made difficult by the scale of the driving processes such as convective instability and vertical wind shear. Although larger-scale systems such as hurricanes and middle-latitude cyclones can be linked to large-scale weather patterns, these smaller events are generated from the interactions between the larger circulation and the localized processes that are below the resolvable scale of global models and, therefore, cannot be directly integrated into factual versus counterfactual modeling scenarios (NASEM 2016). Detection of trends is further challenged by the decrease in days of tornado outbreaks, combined with an increase in the number of tornadoes within these outbreaks (Brooks et al. 2014, Elsner et al. 2015).

Given the limitations of the observational record, the two main approaches to assess historical changes and future occurrences of these events are dynamical modeling or statistical modeling based on proxy large-scale information. Improved computational power has facilitated higher-resolution models that can explicitly resolve the convective processes critical for simulating supercell thunderstorms (Brimelow et al. 2017, Komurcu et al. 2018, Mahoney et al. 2013). This improved modeling capability is important, because projections indicate that convective potential is expected to increase, whereas large-scale shear is expected to decrease (Brooks 2013, Hoogewind et al. 2017, Seeley and Romps 2015), both of which are hard to resolve in large-scale models. Statistical downscaling studies that associate proxy large-scale and more-resolvable features to severe thunderstorm (including hail and tornadoes) have also found that overall potential storm intensity is expected to increase (Diffenbaugh et al. 2013; Gensini et al. 2014; Seeley and Romps 2015; Trapp et al. 2007, 2009; Van Klooster and Roebber 2009) with a corresponding decrease in the frequency of events (Brimelow et al. 2017) but producing larger hail (Prein and Holland 2018) or higher-impact tornadoes (Strader et al. 2017).

2.10 TROPICAL CYCLONES

Generating strong winds, heavy rainfall, and storm surge, TCs are one of the most devastating natural hazards and responsible for large numbers of fatalities and economic losses (e.g., hurricanes Katrina of 2005, Sandy of 2012, Harvey–Irma–Maria of 2017, Florence and Michael of 2018, Dorian of 2019). Increasing evidence shows that TC intensity and precipitation will increase because of climate change (Emanuel 1987, 2013; Hill and Lackmann 2011; Knutson et al. 2010; Webster et al. 2005), although conclusions differ with regard to changes in the overall TC frequency (Bell et al. 2013, Emanuel 2013, Knutson et al. 2010, Murakami et al. 2012). It is also shown that the spatial distribution of TCs may change in the future climate, with decreased TC counts over the southern Gulf of Mexico and Caribbean and increased counts over the central Atlantic (Colbert et al. 2013). Coupled with rapid coastal development and sea-level rise (SLR) (Kopp et al. 2014, Nicholls and Cazenave 2010), future TCs are likely to be more devastating.

Quantifying TC hazards and probabilities in the future climate is challenging. TCs (unlike extratropical cyclones) cannot be well resolved in typical climate models because of their relatively small scales, except perhaps in a few recently developed high-resolution climate models (Murakami et al. 2012, 2015). Dynamic downscaling methods can be used to better resolve TCs in climate-model projections (Knutson et al. 2015), but most of these methods are computationally too expensive to be directly applied to risk analysis. An effective approach is to generate large numbers of synthetic TCs under reanalysis or GCM-projected climate conditions to drive hazard modeling. One widely used TC climatology model that can generate such synthetic TCs was developed by Vickery et al. (2000, 2009), using historical track and climate data to develop a basin-scale full track model and describing TC intensity along each track as a function of the previous step, position, and environmental variables such as sea surface temperature (SST).

The Vickery et al. (2000, 2009) model was applied to develop the design wind maps of ASCE building code, although the potential effect of climate change has not been reflected in the design wind maps. In addition to SST, Yonekura and Hall (2011) and Hall and Yonekura (2013) modeled TC tracks dependent on a large-scale climate variable—El Niño Southern Oscillation (ENSO). Emanuel et al. (2008) and Lee et al. (2018) developed TC climatology models that can generate synthetic TCs driven by more comprehensive climate conditions involving the environmental wind and humidity, thermodynamic state of the atmosphere, and thermal stratification of the ocean.

These TC climatology models have been integrated with TC wind, rainfall, and surge hazard models to investigate how TC hazards may change under climate-model projected future conditions. For example, Mudd et al. (2014) applied the method of Vickery et al. (2009) to study how future SST change will affect the design wind speed and found that the current design wind speed of the US Northeast coastline might be exceeded under future climate scenarios, and that the projected wind speed and rain rate increase at each design level (e.g., 300, 700, and 1,700 year mean return intervals) for the northeastern United States. Ellingwood and Lee (2016) applied a similar method to account for the effect of SST change on hurricane winds; in general, they suggested that time-variant hurricane occurrence and intensity should be adopted in lifecycle reliability assessment of civil infrastructure facilities. Cui and Caracoglia (2016) applied a similar method to study how hurricane winds will change under different climate emission scenarios, also estimating the effect of future warming on structural performance and necessary structural intervention costs. More recently, Xu et al. (2020) applied the TC climatology model of Emanuel et al. (2008) to investigate the evolution of TC wind threat from the past into the future for two vulnerable coastal cities (Hangzhou and Shanghai) in China and discussed its implications for building code specifications. They found that most considered climate models (five out of six) yield significantly larger design wind speeds than the traditional analysis assuming a stationary climate that is currently used in design.

Emanuel (2017) investigated current and future TC rainfall hazards for Houston, Texas, and estimated Hurricane Heavy's (2017) rainfall as having a 0.5% annual probability (1 in a 2,000 year event) in the late twentieth century compared with a 1% annual probability (1 in a 100 year event by the end of this century. Marsooli et al. (2019) applied the TC climatology model of Emanuel et al. (2008) to study potential future changes of storm surge flood along the US Atlantic Coast. They found that, under the compound effects of TC climatology change and SLR over the twenty-first century, the 100 year flood level (1% annual probability) will increase substantially, especially for the Gulf of Mexico region [Figure 2-8(b)]. Similarly, the small range in historical flood levels in higher-latitude regions will result in a relatively quick increase in the annual probability of floods currently assessed at 1% [100 year flood; Figure 2-8(a)]. By the end of the twenty-first century, these flood levels will have a probability of occurring at least once per year in the New England and mid-Atlantic regions and 3% to 100% probability (1 to 30 years) in the southeast Atlantic and Gulf of Mexico regions [Figure 2-8(c)].

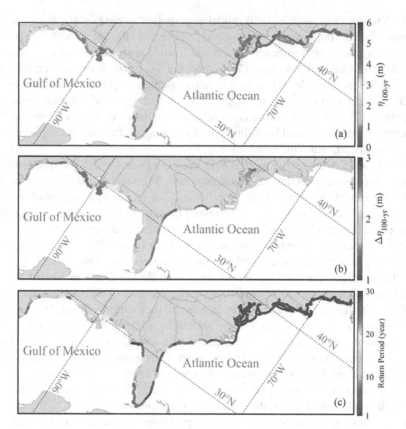

Figure 2-8. Investigation of TC coastal flood hazards for the US Atlantic Coast by Marsooli et al. (2019). Top panel: reanalysis–based estimate of 100 year flood level (1% annual probability) $\eta_{100year}$ for the historical period of 1980 to 2005 (warm color indicates higher risk). Middle panel: projected weighted-average changes in $\eta_{100year}$ for the future period of 2070–2095 (the warm color indicates a larger increase of risk). Bottom panel: future return period of historical 100 year (1% annual probability) flood levels (the cold color indicates a larger increase of risk). Six CMIP5 GCMs are applied. Flood levels are relative to MHHW.

2.11 SEA-LEVEL RISE

Approximately 93% of the excess heat arising from human-induced global warming has been absorbed by the oceans since the mid-twentieth century (USGCRP 2018), resulting in a global increase in SST of 1.2 to 1.4 °F between 1900 and 2016. By 2100, SST is projected to increase by approximately 3.6 to 6.2 F relative to late-twentieth-century SSTs. Such warming leads to thermal expansion of ocean water and thus higher sea levels. Changes in temperature also result in ice melt near the poles and on inland mountain glaciers, releasing a greater volume of freshwater into the oceans (Church et al. 2013, Sweet et al. 2017). The mean global sea level has risen 7 to 8 in. since 1900 from this combination of higher ocean temperatures and increased volume of seawater, with approximately half of this rise occurring since 1993 because of a significant increase in the melting of land-based ice (Hay et al. 2015, Church and White 2011). By 2100, it is very likely that average global sea levels will rise a further 1 to 4 ft because of increases in global temperatures based on the assumption that the historic relationship between global air temperature and global average sea level will continue unchanged (USGCRP 2018). With that said, it should be noted that according to Sweet et al. (2017), the specific emission scenario that is followed through the twenty-first century will not have much impact on SLR through the middle of the century but will have a substantial impact on sea levels during the second half of the century and later (Figure 2-9).

Consistent with average global sea levels, sea levels proximate to the continental United States have been rising, albeit with differing spatial signals between 1853 and 2014 along the West Coast, Gulf of Mexico, and the East Coast (Watson 2016). The average rate of rise in the Central Pacific and along the West Coast was estimated to be 0.05 in./year ± 0.02 in./year, whereas within the Gulf of Mexico and along the East Coast, the average rate was 0.09 ± 0.02 in./year. Such a rise has led to a 5- to 10-fold increase in daily tidal flooding capable of causing minor infrastructure damage since the 1960s in several coastal cities (Sweet et al. 2014). Sweet et al. (2018) also estimate that these minor floods would become major destructive events with an SLR of 1 to 2 ft.

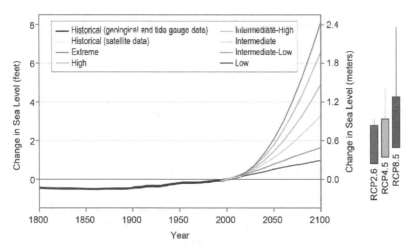

Figure 2-9. Historic and projected changes in global average sea level based on the response of the climate system to warming and the future scenario of the emissions of heat-trapping gases that are employed. The colored lines represent six different scenarios of SLR developed by the US Federal Interagency Sea Level Rise Task Force, whereas the box plots on the right indicate the range of very likely sea levels by the year 2100 based on the three RCP emission scenarios.

Source: Adapted from Sweet et al. (2017).

Lower RSLR along the Northwest Pacific Coast (i.e., Washington, Oregon, and northern California) compared with the global average can be partially attributed to uplift caused by interseismic strain throughout the Cascadia subduction zone (NRC 2012). More recently (post-1980), SLR along the West Coast has been mitigated by a shift in the predominant wind stress regime likely associated with a shift from the cold to warm phases of the PDO (Bromirski et al. 2011). Evidence of a potential return to the cold phase of the PDO may be a harbinger of significant SLR along the Pacific Coast in the near future. In addition to coastal areas, the Sacramento–San Joaquin Delta, which is a vital component of California's water supply system, is particularly vulnerable because of the fact that much land within the Delta is located at or below sea level and protected by an extensive levee network. Increased vulnerability from levee failure and saltwater intrusion owing to increased tide amplitude, storm surges, and rising groundwater presents a major concern for this area (Hanak and Lund 2012). Evidence suggests a recent acceleration (post-2006) in RSLR within the Gulf of Mexico and along the East Coast, although the increase in RSLR is only significant at fewer than 9% of the 2,900 stations analyzed (Watson 2016).

Other factors also contribute to the magnitude and acceleration of SLR along the Gulf and East Coasts, such as vertical land movement from postglacial isostatic readjustment and changes in the strength of the Atlantic meridional overturning circulation (AMOC). The withdrawal of groundwater along both coasts and of fossil fuels along the Gulf Coast have significantly amplified RSLR in these areas (Sweet et al. 2017) compared with the West Coast. RSLR along the East Coast has also been linked to a weakening of the AMOC (Sweet et al. 2017). The narrow Gulf Stream and North Atlantic Currents, which form a portion of the AMOC, create a steep sea-level slope that results in anomalously low sea levels along the East Coast. A collapse of the AMOC, owing to warmer ocean temperatures and lower density of ocean water, would drive higher sea levels along the East Coast. Model simulations performed by Yin et al. (2009) project that the rate of SLR along the northeast coast of the United States will substantially outpace the global mean throughout the remainder of the twenty-first century because of a weaker AMOC.

2.12 OCEAN WAVES

Along the coast, waves represent a major contribution to flooding because of temporary sea-level extremes. Coastal waves also have a major impact on erosion and sediment budgets. Extreme waves located away from the coast can act as hazards for offshore platforms and operations and maritime navigation activities in general. Understanding potential future changes in wave heights is, therefore, a high priority for the sectors previously mentioned.

A warmer climate can have a substantial impact on waves across the globe. Global wave models forced by GCM-simulated near-surface winds have been used to estimate future impacts of a warmer climate on wave heights. For example,

Semedo et al. (2012) found that annual mean wave heights are projected to increase substantially (up to 30%) throughout the Arctic Ocean and along the east coast of Greenland by the end of the twenty-first century compared with the end of the twentieth century. Significant increases of up to 10% are also projected within the Bering Strait and the North Atlantic. In contrast, wave heights are projected to decrease at lower latitudes, particularly along the East Coast of the United States, where decreases of up to 10% are likely.

Seasonal average wave heights are projected to decrease during the boreal winter (December to February) throughout the Northern Hemisphere, with the most extreme changes (>10% change) occurring in the northern Pacific and the central Atlantic Oceans. Higher-wave heights during boreal winter are only likely in the extremely high latitudes and along the east coast of Greenland. Wave heights are also projected to decrease (>10%) during the summer months (June to August) throughout the northern half of the Atlantic Ocean, including the East Coast of the United States, whereas less substantial increases are projected throughout the Pacific Ocean. The greatest increase in wave heights (20% to 30%) is again expected at the extreme high latitudes and along the east coast of Greenland. The general poleward shift of increased wave heights is likely attributable to a poleward shift in midlatitude storm tracks (Mori et al. 2010, Shimura et al. 2015).

Dobrynin et al. (2015) found that climate change signals in terms of changes in wave heights can be observed and distinguished from natural climate variability much earlier than the end of the twenty-first century. Using the output from an earth system model to force a global wave model, detectable changes in wave height owing to climate change were identified as early as within the next two decades for major portions of the northern and equatorial Atlantic Ocean. Throughout much of the rest of the globe, excluding the western and southern Pacific Ocean, a combination of increasing local wind speeds with an additional swell, in addition to increasing SSTs and tropical cyclone activity (Shimura et al. 2015), is expected to lead to an increase in wave heights as early as the current decade.

2.13 WIND SPEED

Future regional changes in wind climate can be estimated using results from GCMs and RCMs, respectively; however, their coarse resolution results in poor representations of the rough topography that has a strong influence on wind speed and direction (Daines et al. 2016). Statistical (Curry et al. 2012, Culver and Monahan 2013) or dynamical downscaling (Pryor et al. 2012, Daines et al. 2016) is typically required to represent such terrain and achieve representative projections of regional winds.

Given the variability in observational records and limited, to date, ability to simulate wind speed with any accuracy, the little research conducted to date has focused on changes in wind characteristics (Jeong and Sushama 2019).

Daines et al. (2016) found that large changes in mean wind speed are unlikely over the provinces of British Columbia and Alberta in the 2031 to 2060 period, whereas Cheng et al. (2012) identified variable increases in the frequency of larger magnitude wind events over Ontario, Canada. On the contrary, Pryor et al. (2012) projected decreases in mean, 90th-, and 95th-percentile wind speeds, particularly in the western United States, for the same time period, although they found no evidence of changes in observed extreme wind speeds in the past century. However, Coburn (2019) cautioned that many of the differences between these assessments arise from the different reanalysis products used to validate GCM and RCM simulations, in addition to structural differences in the models and internal climate variability (Daines et al. 2016). Similarly, Cheng et al. (2012) found that projected increases in the frequency of wind gust events are equal to or exceed uncertainties related to different GCM models.

Projected increases in the severity and frequency of other extreme weather events (e.g., hurricanes, tropical storms, tornadoes, and blizzards) with which extreme winds are associated are expected to result in corresponding increases in the intensity and frequency of wind gusts. Increased temperatures create conditions for stronger winds during hurricanes and tropical storms, and stronger tornadoes owing to increased instability in the lower levels of the atmosphere (Biello 2013).

2.14 COMPOUND AND CASCADING HAZARDS

Droughts, heatwaves, wildfires, and floods often result from interactions between different processes and hazards. The combination of different hazards or processes leading to a significant impact is referred to as a compound event, for example, compound ocean and fluvial flooding in coastal areas (Zscheischler et al. 2018). A cascading hazard is a type of compound event in which consecutive events in a particular order lead to significant impacts (AghaKouchak et al. 2018). For instance, the following set of consecutive events led to significant impacts in California: (1) A prolonged (2012 to 2016) extreme drought increased tree mortality; (2) extreme precipitation from atmospheric rivers in winter 2017 enhanced the growth of fuels such as shrubs and grass; (3) a very dry and warm spring and summer reduced moisture levels and dried the vegetation; (4) Diablo and Santa Ana winds had set records for sustained winds and low humidity; (5) extreme fires occurred shortly after (i.e., the Thomas Fire in December 2017); and (6) a month later, extreme rainfall over burned areas created the most deadly debris flow event in California's history. Unfortunately, traditional risk assessment frameworks cannot be used to describe the risk associated with such compound and cascading events. Most existing methods consider one hazard at a time, potentially leading to an underestimation of the actual risk (Sadegh et al. 2018, Moftakhari et al. 2019). A better understanding of compound and cascading hazards is critical, especially because many drivers of extreme events, including wildfires, extreme rain, and heat waves, are projected to intensify in the future (Perkins 2015, Ragno et al. 2018, USGCRP 2018). This is a major research gap that requires bridging between different disciplines, including climate science, engineering, social science, economics, and policy (Zscheischler et al. 2018).

2.15 CONTEXTUALIZING EXTREME EVENTS WITH RESPECT TO SCALE AND PREDICTABILITY

The spectrum of extreme events discussed in this chapter must be placed in the context of their predictability, sensitivity (and attributability) to climate change, and the relative threat that they pose. As noted in the preceding paragraphs, some severe weather may not be hazardous in isolation, but, as part of a time-evolving compound event, can present far greater risks. Further, the influence of natural variability may result in higher risks in the short term (i.e., decadal) and heighten localized vulnerability more than longer projections indicate. Some of the more devastating aspects of extreme weather are not readily simulated because of their small geographical footprints, tempting decision-makers to postpone adaptation actions for impacted infrastructure. However, research

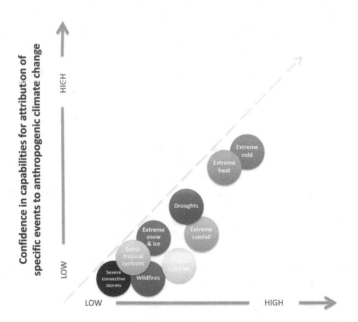

Figure 2-10. Comparison of the types of extreme events presented with respect to the state of the science in assessing the role of climate change in their attribution.
Source: NASEM (2016).

linking those events to large-scale atmospheric dynamics that are skillfully represented can be used as proxy information to guide decisions. To address this challenge, a recent comprehensive report presents the various methods used to assess the influence of climate change on extreme weather (NASEM 2016). In addition to a detailed presentation of methods and current understanding, they synthesize the current knowledge of the effects of climate change on extreme weather with respect to the spatial scale (Figure 2-10). In essence, the larger the spatiotemporal scale of an event, the more likely it can be understood through the lens of climate change and is distinguishable from natural variability. The following chapters link the information presented here to the state of the science, and how it can be used to best prioritize necessary adaptation actions in the short and long term.

References

Abatzoglou, J. T., and A. P. Williams. 2016. "Impact of anthropogenic climate change on wildfire across western US forests." *Proc. Natl. Acad. Sci. U.S.A.* 113 (42): 11770–11775.

AghaKouchak, A., D. Feldman, M. Hoerling, T. Huxman, and J. Lund. 2015. "Water and climate: Recognize anthropogenic drought." *Nature* 524 (7566): 409–411.

AghaKouchak, A., L. S. Huning, F. Chiang, M. Sadegh, F. Vahedifard, O. Mazdiyasni, et al. 2018. "How do natural hazards cascade to cause disasters?" *Nature* 561 (7724): 458–460.

Allan, R. P., and B. J. Soden. 2008. "Atmospheric warming and the amplification of precipitation extremes." *Science* 321 (5895): 1481–1484.

Armal, S., N. Devineni, and R. Khanbilvardi. 2018. "Trends in extreme rainfall frequency in the contiguous United States: Attribution to climate change and climate variability modes." *J. Clim.* 31 (1): 369–385.

ASCE CACC (Committee on Adaptation to a Changing Climate). 2015. *Adapting infrastructure and civil engineering practice to a changing climate*, edited by J. R. Rolf. Reston, VA: ASCE.

Barnett, T. P., J. C. Adam, and D. P. Lettenmaier. 2005. "Potential impacts of a warming climate on water availability in snow-dominated regions." *Nature* 438 (7066): 303–309.

Barnston, A. G., and R. E. Livezey. 1987. "Classification, seasonality and persistence of low-frequency atmospheric circulation patterns." *Mon. Weather Rev.* 115 (6): 1083–1126.

Bell, R., J. Strachan, P. L. Vidale, K. Hodges, and M. Roberts. 2013. "Response of tropical cyclones to idealized climate change experiments in a global high-resolution coupled general circulation model." *J. Clim.* 26 (20): 7966–7980.

Bender, F. A.-M., V. Ramanathan, and G. Tselioudis. 2012. "Changes in extratropical storm track cloudiness 1983–2008: Observational support for a poleward shift." *Clim. Dyn.* 38 (9–10): 2037–2053.

Bentz, B. J., J. Régnière, C. J. Fettig, E. M. Hansen, J. L. Hayes, J. A. Hicke, et al. 2010. "Climate change and bark beetles of the western United States and Canada: Direct and indirect effects." *BioScience* 60 (8): 602–613.

Bertrand, D., and R. A. McPherson. 2018. "Future hydrologic extremes of the Red River Basin." *J. Appl. Meteorol. Climatol.* 57 (6): 1321–1336.

Biello, D. 2013. "What role does climate change play in tornadoes?" *Scientific American May 21, 2013*. Accessed January 15, 2019. https://www.scientificamerican.com/article/kevin-trenberth-on-climate-change-and-tornadoes/.

Bonnin, G. M., K. Maitaria, and M. Yekta. 2011. "Trends in rainfall exceedances in the observed record in selected areas of the United States." *JAWRA J. Am. Water Resour. Assoc.* 47 (6): 1173–1182.

Brimelow, J. C., W. R. Burrows, and J. M. Hanesiak. 2017. "The changing hail threat over North America in response to anthropogenic climate change." *Nat. Clim. Change* 7 (7): 516–522.

Bromirski, P. D., A. J. Miller, R. E. Flick, and G. Auad. 2011. "Dynamical suppression of sea level rise along the Pacific coast of North America: Indications for imminent acceleration." *J. Geophys. Res.* 116: C07005.

Brooks, B. A., G. Bawden, D. Manjunath, C. Werner, N. Knowles, J. Foster, et al. 2012. "Contemporaneous subsidence and levee overtopping potential, Sacramento–San Joaquin Delta, California." *San Francisco Estuary Watershed Sci.* 10 (1): 1–18.

Brooks, H. E. 2013. "Severe thunderstorms and climate change." *Atmos. Res.* 123: 129–138.

Brooks, H. E., G. W. Carbin, and P. T. Marsh. 2014. "Increased variability of tornado occurrence in the United States." *Science* 346 (6207): 349–352.

Cannon, S. H. 2001. "Debris-flow generation from recently burned watersheds." *Environ. Eng. Geosci.* 7 (4): 321–341.

Chan, S. C., E. J. Kendon, N. Roberts, S. Blenkinsop, and H. J. Fowler. 2018. "Large-scale predictors for extreme hourly precipitation events in convection-permitting climate simulations." *J. Clim.* 31 (6): 2115–2131.

Chang, E. K. M. 2013. "CMIP5 projection of significant reduction in extratropical cyclone activity over North America." *J. Clim.* 26 (24): 9903–9922.

Chen, X., P. Wu, M. J. Roberts, and T. Zhou. 2018. "Potential underestimation of future Mei-Yu rainfall with coarse-resolution climate models." *J. Clim.* 31 (17): 6711–6727.

Cheng, C. S., G. Li, Q. Li, H. Auld, and C. Fu. 2012. "Possible impacts of climate change on wind gusts under downscaled future climate conditions over Ontario, Canada." *J. Clim.* 25 (9): 3390–3408.

Cheng, L., M. Hoerling, L. Smith, and J. Eischeid. 2018. "Diagnosing human-induced dynamic and thermodynamic drivers of extreme rainfall." *J. Clim.* 31 (3): 1029–1051.

Chiang, F., O. Mazdiyasni, and A. AghaKouchak. 2018. "Amplified warming of droughts in Southern United States in observations and model simulations." *Sci. Adv.* 4 (8): eaat2380.

Church, J. A., and N. J. White. 2011. "Sea-level rise from the late 19th to the early 21st century." *Surv. Geophys.* 32 (4–5): 585–602.

Church, J. A., et al. 2013. "Sea level change." In *Climate Change 2013: The Physical Science Basis. Contribution of Working Group I to the Fifth Assessment Report of the Intergovernmental Panel on Climate Change*, edited by T. F. Stocker, D. Qin, G.-K. Plattner, M. Tignor, S. K. Allen, J. Boschung, et al., 1137–1216. Cambridge, UK: Cambridge University Press.

Clabo, D. R. 2018. "Contemporary pyrogeography and wildfire–climate relationships of South Dakota, USA." *Atmosphere* 9 (6): 207.

Coats, S., B. I. Cook, J. E. Smerdon, and R. Seager. 2015. "North American pancontinental droughts in model simulations of the last millennium." *J. Clim.* 28 (5): 2025–2043.

Coburn, J. J. 2019. "Assessing wind data from reanalyses for the upper midwest." *J. Appl. Meteorol. Climatol.* 58 (3): 429–446.

Colbert, A. J., B. J. Soden, G. A. Vecchi, and B. P. Kirtman. 2013. "The impact of anthropogenic climate change on North Atlantic tropical cyclone tracks." *J. Clim.* 26 (12): 4088–4095.

Colle, B. A., Z. Zhang, K. A. Lombardo, E. Chang, P. Liu, and M. Zhang. 2013. "Historical evaluation and future prediction of eastern North American and western Atlantic extratropical cyclones in the CMIP5 models during the cool season." *J. Clim.* 26 (18): 6882–6903.

Collins, M., et al. 2013. "Long-term climate change: Projections, commitments and irreversibility." In *Climate Change 2013: The Physical Science Basis. Contribution of Working Group I to the Fifth Assessment Report of the Intergovernmental Panel on Climate Change,* edited by T. F. Stocker, D. Qin, G.-K. Plattner, M. Tignor, S. K. Allen, J. Boschung, et al., 1029–1136. Cambridge, UK: Cambridge University Press.

Crockett, J. L., and A. L. Westerling. 2018. "Greater temperature and precipitation extremes intensify western U.S. droughts, wildfire severity, and Sierra Nevada tree mortality." *J. Clim.* 31 (1): 341–354.

Cui, W., and L. Caracoglia. 2016. "Exploring hurricane wind speed along US Atlantic coast in warming climate and effects on predictions of structural damage and intervention costs." *Eng. Struct.* 122: 209–225.

Culver, A. M. R., and A. H. Monahan. 2013. "The statistical predictability of surface winds over western and central Canada." *J. Clim.* 26 (21): 8305–8322.

Curry, C. L., D. van der Kamp, and A. H. Monahan. 2012. "Statistical downscaling of historical monthly mean winds over a coastal region of complex terrain. I. Predicting wind speed." *Clim. Dyn.* 38 (7–8): 1281–1299.

Curtis, S., and R. Adler. 2000. "ENSO indices based on patterns of satellite-derived precipitation." *J. Clim.* 13 (15): 2786–2793.

Dai, A. 2013. "Increasing drought under global warming in observations and models." *Nat. Clim. Change* 3 (1): 52–58.

Daines, J. T., A. H. Monahan, and C. L. Curry. 2016. "Model-based projections and uncertainties of near-surface wind climate in western Canada." *J. Appl. Meteorol. Climatol.* 55 (10): 2229–2245.

Dennison, P. E., S. C. Brewer, J. D. Arnold, and M. A. Moritz. 2014. "Large wildfire trends in the western United States, 1984–2011." *Geophys. Res. Lett.* 41 (8): 2928–2933.

Dettinger, M. 2005. *Changes in streamflow timing in the western United States in recent decades.* USGS Fact Sheet 2005-3018. Reston, VA: US Geological Survey.

Dettinger, M. 2016. "Historical and future relations between large storms and droughts in California." *San Francisco Estuary Watershed Sci.* 14 (2): 1–21.

Diffenbaugh, N. S., M. Scherer, and R. J. Trapp. 2013. "Robust increases in severe thunderstorm environments in response to greenhouse forcing." *Proc. Natl. Acad. Sci. U.S.A.* 110 (41): 16361–16366.

Dobrynin, M., J. Murawski, J. Baehr, and T. Ilyina. 2015. "Detection and attribution of climate change signal in ocean wind waves." *J. Clim.* 28 (4): 1578–1591.

Done, J. M., G. J. Holland, C. L. Bruyère, L. R. Leung, and A. Suzuki-Parker. 2015. "Modeling high-impact weather and climate: Lessons from a tropical cyclone perspective." *Clim. Change* 129 (3–4): 381–395.

Easterling, D. R., K. E. Kunkel, J. R. Arnold, T. Knutson, A. N. LeGrande, L. R. Leung, et al. 2017. "Precipitation change in the United States." In Vol. 1 of *Climate Science Special Report: Fourth National Climate Assessment,* edited by D. J. Wuebbles, D. W. Fahey, K. A. Hibbard, D. J. Dokken, B. C. Stewart, and T. K. Maycock, 207–230. Washington, DC: United States Global Change Research Program.

Ellingwood, B. R., and J. Y. Lee. 2016. "Life cycle performance goals for civil infrastructure: Intergenerational risk-informed decisions." *Struct. Infrastruct. Eng.* 12 (7): 822–829.

Elsner, J. B., S. C. Elsner, and T. H. Jagger. 2015. "The increasing efficiency of tornado days in the United States." *Clim. Dyn.* 45 (3–4): 651–659.

Emanuel, K. A. 1987. "The dependence of hurricane intensity on climate." *Nature* 326 (6112): 483–485.

Emanuel, K. A. 2013. "Downscaling CMIP5 climate models shows increased tropical cyclone activity over the 21st century." *Proc. Natl. Acad. Sci. U.S.A.* 110 (30): 12219–12224.

Emanuel, K. 2017. "Assessing the present and future probability of Hurricane Harvey's rainfall." *Proc. Natl. Acad. Sci. U.S.A.* 114 (48): 12681–12684.

Emanuel, K. A., R. Sundararajan, and J. Williams. 2008. "Hurricanes and global warming: Results from downscaling IPCC AR4 simulations." *Bull. Am. Meteorol. Soc.* 89 (3): 347–368.

Farnham, D. J., J. Doss-Gollin, and U. Lall. 2018. "Regional extreme precipitation events: Robust inference from credibly simulated GCM variables." *Water Resour. Res.* 54 (6): 3809–3824.

Finney, D. L., R. M. Doherty, O. Wild, D. S. Stevenson, I. A. MacKenzie, and A. M. Blyth. 2018. "A projected decrease in lightning under climate change." *Nat. Clim. Change* 8 (3): 210–213.

Flannigan, M., B. Stocks, M. Turetsky, and M. Wotton. 2009. "Impacts of climate change on fire activity and fire management in the circumboreal forest." *Global Change Biol.* 15 (3): 549–560.

Francis, J. A., and S. J. Vavrus. 2012. "Evidence linking Arctic amplification to extreme weather in mid-latitudes." *Geophys. Res. Lett.* 39 (6): L06801.

Fritze, H., I. T. Stewart, and E. Pebesma. 2011. "Shifts in western North American snowmelt runoff regimes for the recent warm decades." *J. Hydrometeorol.* 12 (5): 989–1006.

Fuchs, B., D. Wood, and D. Ebbeka. 2015. *From Too Much to Too Little: How the central U.S. drought of 2012 evolved out of one of the most devastating floods on record in 2011.* Lincoln, NE: Drought Mitigation Center Faculty Publications.

Gao, X., C. A. Schlosser, P. A. O'Gorman, E. Monier, and D. Entekhabi. 2017. "Twenty-first-century changes in U.S. regional heavy precipitation frequency based on resolved atmospheric patterns." *J. Clim.* 30 (7): 2501–2521.

Gensini, V. A., C. Ramseyer, and T. L. Mote. 2014. "Future convective environments using NARCCAP." *Int. J. Climatol.* 34 (5): 1699–1705.

Gershunov, A., and K. Guirguis. 2012. "California heat waves in the present and future." *Geophys. Res. Lett.* 39 (18): L18710.

Gershunov, A., T. Shulgina, R. E. S. Clemesha, K. Guirguis, D. W. Pierce, M. D. Dettinger, et al. 2019. "Precipitation regime change in Western North America: The role of atmospheric rivers." *Sci. Rep.* 9 (1): 9944.

Giovannettone, J. 2021. "Assessing the relationship between low-frequency oscillations of global hydro-climate indices and long-term precipitation throughout the United States." *J. Appl. Meteorol. Climatol.* 60 (1): 87–101.

Giovannettone, J., and Y. Zhang. 2019. "Identifying strong signals between low-frequency climate oscillations and annual precipitation using correlation analysis." *Int. J. Climatol.* 39 (12): 4883–4894.

Guzman-Morales, J., and A. Gershunov. 2019. "Climate change suppresses Santa Ana Winds of southern California and sharpens their seasonality." *Geophys. Res. Lett.* 46 (5): 2772–2780.

Hall, T. M., and E. Yonekura. 2013. "North American tropical cyclone landfall and SST: A statistical model study." *J. Clim.* 26 (21): 8422–8439.

Hanak, E., and J. R. Lund. 2012. "Adapting California's water management to climate change." *Clim. Change* 111 (1): 17–44.

Harpold, A. A., M. Dettinger, and S. Rajagopal. 2017. "Defining snow drought and why it matters." *Eos, Earth Space Sci. News* 98 (5): 15–17.

Hatchett, B. J., D. P. Boyle, A. E. Putnam, and S. D. Bassett. 2015. "Placing the 2012–2015 California–Nevada drought into a paleoclimatic context: Insights from walker lake, California–Nevada, USA." *Geophys. Res. Lett.* 42 (20): 8632–8640.

Hay, C. C., E. Morrow, R. E. Kopp, and J. X. Mitrovica. 2015. "Probabilistic reanalysis of twentieth-century sea-level rise." *Nature* 517 (7535): 481–484.

Hayhoe, K., D. J. Wuebbles, D. R. Easterling, D. W. Fahey, S. Doherty, J. Kossin, et al. 2018. "Our changing climate." In Vol. 2 of *Impacts, Risks, and Adaptation in the United States: Fourth National Climate Assessment*, edited by D. R. Reidmiller, C. W. Avery, D. R. Easterling, K. E. Kunkel, K. L. M. Lewis, T. K. Maycock, et al., 72–144. Washington, DC: US Global Change Research Program.

Herrera-Estrada, J. E., and J. Sheffield. 2017. "Uncertainties in future projections of summer droughts and heat waves over the contiguous United States." *J. Clim.* 30 (16): 6225–6246.

Herring, S. C., N. Christidis, A. Hoell, M. P. Hoerling, and P. A. Stott. 2019. "Explaining extreme events of 2017 from a climate perspective." *Bull. Am. Meteorol. Soc.* 100 (1): S1–S117.

Hibbard, K. A., F. M. Hoffman, D. Huntzinger, and T. O. West. 2017. "Changes in land cover and terrestrial biogeochemistry." In Vol. 1 of *Climate Science Special Report: Fourth National Climate Assessment*, edited by D. J. Wuebbles, D. W. Fahey, K. A. Hibbard, D. J. Dokken, B. C. Stewart, and T. K. Maycock, 277–302. Washington, DC: US Global Change Research Program.

Hill, K. A., and G. M. Lackmann. 2011. "The impact of future climate change on TC intensity and structure: A downscaling approach." *J. Clim.* 24 (17): 4644–4661.

Hoogewind, K. A., M. E. Baldwin, and R. J. Trapp. 2017. "The impact of climate change on hazardous convective weather in the United States: Insight from high-resolution dynamical downscaling." *J. Clim.* 30 (24): 10081–10100.

Huang, X., and P. A. Ullrich. 2017. "The changing character of twenty-first-century precipitation over the western United States in the variable-resolution CESM." *J. Clim.* 30 (18): 7555–7575.

Hurrell, J. W. 1995. "Decadal trends in the North Atlantic oscillation: Regional temperatures and precipitation." *Science* 269 (5224): 676–679.

IPCC (Intergovernmental Panel on Climate Change). 2013. "Summary for policymakers." In *Climate Change 2013: The Physical Science Basis. Contribution of Working Group I to the Fifth Assessment Report of the Intergovernmental Panel on Climate Change*, edited by T. F. Stocker, D. Qin, G.-K. Plattner, M. Tignor, S. K. Allen, J. Boschung, et al., 3–29. Cambridge, UK: Cambridge University Press.

Jenkins, M. J., J. B. Runyon, C. J. Fettig, W. G. Page, and B. J. Bentz. 2014. "Interactions among the mountain pine beetle, fires, and fuels." *For. Sci.* 60 (3): 489–501.

Jeong, D. I., and L. Sushama. 2019. "Projected changes to mean and extreme surface wind speeds for North America based on regional climate model simulations." *Atmosphere* 10 (9): 497.

Johnson, N. C., S.-P. Xie, Y. Kosaka, and X. Li. 2018. "Increasing occurrence of cold and warm extremes during the recent global warming slowdown." *Nat. Commun.* 9 (1): 1724.

Jolly, W. M., R. A. Parsons, A. M. Hadlow, G. M. Cohn, S. S. McAllister, J. B. Popp, et al. 2012. "Relationships between moisture, chemistry, and ignition of *Pinus contorta* needles during the early stages of mountain pine beetle attack." *For. Ecol. Manage.* 269: 52–59.

Joly, K., P. A. Duffy, and T. S. Rupp. 2012. "Simulating the effects of climate change on fire regimes in Arctic biomes: Implications for caribou and moose habitat." *Ecosphere* 3 (5): 1–18.

Kaufmann, R. K., H. Kauppi, and J. H. Stock. 2010. "Does temperature contain a stochastic trend? Evaluating conflicting statistical results." *Clim. Change* 101 (3–4): 395–405.

Keellings, D., and H. Moradkhani. 2020. "Spatiotemporal evolution of heat wave severity and coverage across the United States." *Geophys. Res. Lett.* 47 (9): e2020GL087097.

Kendon, E. J., S. Blenkinsop, and H. J. Fowler. 2018. "When will we detect changes in short-duration precipitation extremes?" *J. Clim.* 31 (7): 2945–2964.

Knowles, N., M. D. Dettinger, and D. R. Cayan. 2006. "Trends in snowfall versus rainfall in the western United States." *J. Clim.* 19 (18): 4545–4559.

Knutson, T. R., J. L. McBride, J. Chan, K. Emanuel, G. Holland, C. Landsea, et al. 2010. "Tropical cyclones and climate change." *Nat. Geosci.* 3 (3): 157–163.

Knutson, T. R., J. J. Sirutis, M. Zhao, R. E. Tuleya, M. Bender, G. A. Vecchi, et al. 2015. "Global projections of intense tropical cyclone activity for the late twenty-first century from dynamical downscaling of CMIP5/RCP4.5 scenarios." *J. Clim.* 28 (18): 7203–7224.

Komurcu, M., K. A. Emanuel, M. Huber, and R. P. Acosta. 2018. "High-resolution climate projections for the northeastern United States using dynamical downscaling at convection-permitting scales." *Earth Space Sci.* 5 (11): 801–826.

Kopp, R. E., R. M. Horton, C. M. Little, J. X. Mitrovica, M. Oppenheimer, D. J. Rasmussen, et al. 2014. "Probabilistic 21st and 22nd century sea-level projections at a global network of tide-gauge sites." *Earth's Future* 2 (8): 383–406.

Kug, J., J.-H. Jeong, Y.-S. Jang, B.-M. Kim, C. K. Folland, S.-K. Min, et al. 2015. "Two distinct influences of Arctic warming on cold winters over North America and East Asia." *Nat. Geosci.* 8 (10): 759–762.

Lee, C.-Y., M. K. Tippett, A. H. Sobel, and S. J. Camargo. 2018. "An environmentally forced tropical cyclone hazard model." *J. Adv. Model. Earth Syst.* 10 (1): 223–241.

Lehmann, J., D. Coumou, and K. Frieler. 2015. "Increased record-breaking precipitation events under global warming." *Clim. Change* 132 (4): 501–515.

Li, J., H. Chen, X. Rong, J. Su, Y. Xin, K. Furtado, et al. 2018. "How well can a climate model simulate an extreme precipitation event: A case study using the transpose-AMIP experiment." *J. Clim.* 31 (16): 6543–6556.

Li, J., D. W. J. Thompson, E. A. Barnes, and S. Solomon. 2017. "Quantifying the lead time required for a linear trend to emerge from natural climate variability." *J. Clim.* 30 (24): 10179–10191.

Li, Z., and H. Fang. 2016. "Impacts of climate change on water erosion: A review." *Earth Sci. Rev.* 163: 94–117.

Littell, J. S., D. McKenzie, D. L. Peterson, and A. L. Westerling. 2009. "Climate and wildfire area burned in western U.S. ecoprovinces, 1916–2003." *Ecol. Appl.* 19 (4): 1003–1021.

Littell, J. S., E. E. Oneil, D. McKenzie, J. A. Hicke, J. A. Lutz, R. A. Norheim, et al. 2010. "Forest ecosystems, disturbance, and climatic change in Washington State, USA." *Clim. Change* 102 (1–2): 129–158.

Liu, Y.-C., P. Di, S.-H. Chen, and J. DaMassa. 2018. "Relationships of rainy season precipitation and temperature to climate indices in California: Long-term variability and extreme events." *J. Clim.* 31 (5): 1921–1942.

Liu, Y., S. L. Goodrick, and J. A. Stanturf. 2013. "Future U.S. wildfire potential trends projected using a dynamically downscaled climate change scenario." *For. Ecol. Manage.* 294: 120–135.

Ma, C.-G., and E. K. M. Chang. 2017. "Impacts of storm-track variations on wintertime extreme weather events over the continental United States." *J. Clim.* 30 (12): 4601–4624.

Ma, S., T. Zhou, O. Angélil, and H. Shiogama. 2017. "Increased chances of drought in southeastern periphery of the Tibetan Plateau induced by anthropogenic warming." *J. Clim.* 30 (16): 6543–6560.

MacDonald, R. J., J. M. Byrne, S. W. Kienzle, and R. P. Larson. 2011. "Assessing the potential impacts of climate change on mountain snowpack in the St. Mary River Watershed, Montana." *J. Hydrometeorol.* 12 (2): 262–273.

Madden, R. A., and P. R. Julian. 1994. "Observations of the 40–50-day tropical oscillation—A review." *Mon. Weather Rev.* 122 (5): 814–837.

Mahoney, K., M. Alexander, J. D. Scott, and J. Barsugli. 2013. "High-resolution downscaled simulations of warm-season extreme precipitation events in the Colorado front range under past and future climates." *J. Clim.* 26 (21): 8671–8689.

Marsooli, R., N. Lin, K. Emanuel, and K. Feng. 2019. "Climate change exacerbates hurricane flood hazards along US Atlantic and Gulf Coasts in spatially varying patterns." *Nat. Commun.* 10 (1): 3785.

Martel, J.-L., A. Mailhot, F. Brissette, and D. Caya. 2018. "Role of natural climate variability in the detection of anthropogenic climate change signal for mean and extreme precipitation at local and regional scales." *J. Clim.* 31 (11): 4241–4263.

Martinuzzi, S., S. I. Stewart, D. P. Helmers, M. H. Mockrin, R. B. Hammer, and V. C. Radeloff. 2015. *The 2010 wildland–urban interface of the conterminous United States.* Research Map NRS-8. Newtown Square, PA: US Dept. of Agriculture, Forest Service, Northern Research Station.

Mastin, M. C., K. J. Chase, and R. W. Dudley. 2011. "Changes in spring snowpack for selected basins in the United States for different climate-change scenarios." *Earth Interact* 15 (23): 1–18.

Mazdiyasni, O., and A. AghaKouchak. 2015. "Substantial increase in concurrent droughts and heatwaves in the United States." *Proc. Natl. Acad. Sci. U.S.A.* 112 (37): 11484–11489.

McKenzie, D., Z. E. Gedalof, D. L. Peterson, and P. Mote. 2004. "Climatic change, wildfire, and conservation." *Conserv. Biol.* 18 (4): 890–902.

McKenzie, D., and J. S. Littell. 2017. "Climate change and the eco-hydrology of fire: Will area burned increase in a warming western USA?" *Ecol. Appl.* 27 (1): 26–36.

McKinnon, K. A., and C. Deser. 2018. "Internal variability and regional climate trends in an observational large ensemble." *J. Clim.* 31 (17): 6783–6802.

Melvin, A. M., P. Larsen, B. Boehlert, J. E. Neumann, P. Chinowsky, X. Espinet, et al. 2017. "Climate change damages to Alaska public infrastructure and the economics of proactive adaptation." *Proc. Natl. Acad. Sci. U.S.A.* 114 (2): E122–E131.

Milly, P. C. D., J. Betancourt, M. Falkenmark, R. M. Hirsch, Z. W. Kundzewicz, D. P. Lettenmaier, et al. 2008. "Stationarity is dead: Whither water management?" *Science* 319 (5863): 573–574.

Min, S.-K., X. Zhang, F. W. Zwiers, and G. C. Hegerl. 2011. "Human contribution to more-intense precipitation extremes." *Nature* 470 (7334): 378–381.

Mishra, V., K. A. Cherkauer, and S. Shukla. 2010. "Assessment of drought due to historic climate variability and projected future climate change in the midwestern United States." *J. Hydrometeorol.* 11 (1): 46–68.

Mo, K. C., and D. P. Lettenmaier. 2018. "Drought variability and trends over the central United States in the instrumental record." *J. Hydrometeorol.* 19 (7): 1149–1166.

Moftakhari, H., J. E. Schubert, A. AghaKouchak, R. A. Matthew, and B. F. Sanders. 2019. "Linking statistical and hydrodynamic modeling for compound flood hazard assessment in tidal channels and estuaries." *Adv. Water Resour.* 128: 28–38.

Moody, J. A., and D. A. Martin. 2001. "Post-fire, rainfall intensity-peak discharge relations for three mountainous watersheds in the western USA." *Hydrol. Processes* 15 (15): 2981–2993.

Mori, N., T. Yasuda, H. Mase, T. Tom, and Y. Oku. 2010. "Projection of extreme wave climate change under global warming." *Hydrol. Res. Lett.* 4: 15–19.

Mote, P. W., A. F. Hamlet, M. P. Clark, and D. P. Lettenmaier. 2005. "Declining mountain snowpack in western North America." *Bull. Am. Meteorol. Soc.* 86 (1): 39–50.

Mudd, L., Y. Wang, C. Letchford, and D. Rosowsky. 2014. "Hurricane wind hazard assessment for a rapidly warming climate scenario." *J. Wind Eng.* 133: 242–249.

Murakami, H., G. A. Vecchi, S. Underwood, T. L. Delworth, A. T. Wittenberg, W. G. Anderson, et al. 2015. "Simulation and prediction of category 4 and 5 hurricanes in the high-resolution GFDL HiFLOR coupled climate model." *J. Clim.* 28 (23): 9058–9079.

Murakami, H., Y. Wang, H. Yoshimura, R. Mizuta, M. Sugi, E. Shindo, et al. 2012. "Future changes in tropical cyclone activity projected by the new high-resolution MRI-AGCM." *J. Clim.* 25 (9): 3237–3260.

NASEM (National Academies of Sciences, Engineering, and Medicine). 2016. *Attribution of extreme weather events in the context of climate change.* Washington, DC: National Academy of Sciences, Courtesy of the National Academies Press.

Nearing, M. A., F. F. Pruski, and M. R. O'Neal. 2004. "Expected climate change impacts on soil erosion rates: A review." *J. Soil Water Conserv.* 59 (1): 43–50.

Nicholls, R. J., and A. Cazenave. 2010. "Sea-level rise and its impact on coastal zones." *Science* 328 (5985): 1517–1520.

NOAA NCEI (National Centers for Environmental Information). 2017. "Billion-dollar weather and climate disasters: Overview." Accessed September 30, 2018. https://ncdc. noaa.gov/billions/.

NRC (National Research Council). 2012. "Sea-level variability and change off the California, Oregon, and Washington Coasts." In *Sea-Level Rise for the Coasts of California, Oregon, and Washington: Past, Present, and Future, 55–82.* Washington, DC: National Academies Press.

O'Gorman, P. A. 2015. "Precipitation extremes under climate change." *Curr. Clim. Change Rep.* 1 (2): 49–59.

Otkin, J. A., M. Svoboda, E. D. Hunt, T. W. Ford, M. C. Anderson, C. Hain, et al. 2018. "Flash droughts: A review and assessment of the challenges imposed by rapid-onset droughts in the United States." *Bull. Am. Meteorol. Soc.* 99 (5): 911–919.

Page, W. G., M. J. Jenkins, and J. B. Runyon. 2012. "Mountain pine beetle attack alters the chemistry and flammability of lodgepole pine foliage." *Can. J. For. Res.* 42 (8): 1631–1647.

PaiMazumder, D., and J. M. Done. 2016. "Potential predictability sources of the 2012 U.S. drought in observations and a regional model ensemble." *J. Geophys. Res. Atmos.* 121 (21): 12581–12592.

Partain, J. L., Jr., S. Alden, U. S. Bhatt, P. A. Bieniek, B. R. Brettschneider, R. T. Lader, et al. 2016. "An assessment of the role of anthropogenic climate change in the Alaska fire season of 2015." In *Explaining Extreme Events of 2015 from a Climate Perspective, Bull. Am. Meteorol. Soc.* 97 (12), S14–S18.

Pendergrass, A. G., and D. L. Hartmann. 2014. "Changes in the distribution of rain frequency and intensity in response to global warming." *J. Clim.* 27 (22): 8372–8383.

Perkins, S. E. 2015. "A review on the scientific understanding of heatwaves—Their measurement, driving mechanisms, and changes at the global scale." *Atmos. Res.* 164–165: 242–267.

Prein, A. F., and G. J. Holland. 2018. "Global estimates of damaging hail hazard." *Weather Clim. Extremes* 22: 10–23.

Prein, A. F., G. J. Holland, R. M. Rasmussen, M. P. Clark, and M. R. Tye. 2016. "Running dry: The U.S. Southwest's drift into a drier climate state." *Geophys. Res. Lett.* 43 (3): 1272–1279.

Prein, A. F., G. J. Holland, R. M. Rasmussen, J. Done, K. Ikeda, M. P. Clark, et al. 2013. "Importance of regional climate model grid spacing for the simulation of heavy precipitation in the Colorado headwaters." *J. Clim.* 26 (13): 4848–4857.

Prein, A. F., W. Langhans, G. Fosser, A. Ferrone, N. Ban, K. Goergen, et al. 2015. "A review on regional convection-permitting climate modeling: Demonstrations, prospects, and challenges." *Rev. Geophys.* 53 (2): 323–361.

Prein, A. F., R. M. Rasmussen, K. Ikeda, C. Liu, M. P. Clark, and G. J. Holland. 2017. "The future intensification of hourly precipitation extremes." *Nat. Clim. Change* 7 (1): 48–52.

Prestemon, J. P., U. Shankar, A. Xiu, K. Talgo, D. Yang, E. Dixon, et al. 2016. "Projecting wildfire area burned in the south-eastern United States, 2011–60." *Int. J. Wildland Fire* 25 (7): 715–729.

Pryor, S. C., R. J. Barthelmie, and J. T. Schoof. 2012. "Past and future wind climates over the contiguous USA based on the North American Regional Climate Change Assessment Program model suite." *J. Geophys. Res.* 117: D19119.

Radeloff, V. C., R. B. Hammer, S. I. Stewart, J. S. Fried, S. S. Holcomb, and J. F. McKeefry. 2005. "The wildland–urban interface in the United States." *Ecol. Appl.* 15 (3): 799–805.

Radeloff, V. C., D. P. Helmers, H. A. Kramer, M. H. Mockrin, P. M. Alexandre, A. Bar-Massada, et al. 2018. "Rapid growth of the US wildland–urban interface raises wildfire risk." *Proc. Natl. Acad. Sci. U.S.A.* 115 (13): 3314–3319.

Raei, E., M. R. Nikoo, A. AghaKouchak, O. Mazdiyasni, and M. Sadegh. 2018. "GHWR, A multi-method global heatwave and warm-spell record and toolbox." *Sci. Data* 119: 188–196.

Ragno, E., A. AghaKouchak, C. A. Love, L. Cheng, F. Vahedifard, and C. H. R. Lima. 2018. "Quantifying changes in future intensity-duration-frequency curves using multimodel ensemble simulations." *Water Resour. Res.* 54 (3): 1751–1764.

Riley, K. L., and R. A. Loehman. 2016. "Mid-21st-century climate changes increase predicted fire occurrence and fire season length, Northern Rocky Mountains, United States." *Ecosphere* 7 (11): e01543.

Romps, D. M., J. T. Seeley, D. Vollaro, and J. Molinari. 2014. "Projected increase in lightning strikes in the United States due to global warming." *Science* 346 (6211): 851–854.

Ryan, K. C. 2000. "Global change and wildland fire." In Vol. 2 of *Wildland fire in ecosystems: Effects of fire on flora*, edited by J. K. Brown and J. K. Smith, 175–183. General Technical Report RMRS-GTR-42. Newtown Square, PA: US Dept. of Agriculture, Forest Service, Rocky Mountain Research Station.

Sadegh, M., H. Moftakhari, H. V. Gupta, E. Ragno, O. Mazdiyasni, B. Sanders, et al. 2018. "Multihazard scenarios for analysis of compound extreme events." *Geophys. Res. Lett.* 45 (11): 5470–5480.

Schubert, S., Y. Chang, H. Wang, R. Koster, and M. Suarez. 2016. "A modeling study of the causes and predictability of the spring 2011 extreme U.S. weather activity." *J. Clim.* 29 (21): 7869–7887.

Screen, J. A., and I. Simmonds. 2014. "Amplified mid-latitude planetary waves favour particular regional weather extremes." *Nat. Clim. Change* 4 (8): 704–709.

Seager, R., A. Tzanova, and J. Nakamura. 2009. "Drought in the southeastern United States: Causes, variability over the last millennium, and the potential for future hydroclimate change." *J. Clim.* 22 (19): 5021–5045.

Seeley, J. T., and D. M. Romps. 2015. "The effect of global warming on severe thunderstorms in the United States." *J. Clim.* 28 (6): 2443–2458.

Segura, C., G. Sun, S. McNulty, and Y. Zhang. 2014. "Potential impacts of climate change on soil erosion vulnerability across the conterminous United States." *J. Soil Water Conserv.* 69 (2): 171–181.

Semedo, A., R. Weisse, A. Behrens, A. Sterl, L. Bengtsson, and H. Günther. 2012. "Projection of global wave climate change toward the end of the twenty-first century." *J. Clim.* 26 (21): 8269–8288.

Semmler, T., T. Jung, M. A. Kasper, and S. Serrar. 2018. "Using NWP to assess the influence of the Arctic atmosphere on midlatitude weather and climate." *Adv. Atmos. Sci.* 35 (1): 5–13.

Shao, D., S. Chen, X. Tan, and W. Gu. 2018. "Drought characteristics over China during 1980-2015." *Int. J. Climatol.* 38 (9): 3532–3545.

Sharma, C., C. S. P. Ojha, A. K. Shukla, Q. B. Pham, N. T. T. Linh, C. M. Fai, et al. 2019. "Modified approach to reduce GCM bias in downscaled precipitation: A study in Ganga River Basin." *Water* 11 (10): 2097–2127.

Sheffield, J., E. F. Wood, and M. L. Roderick. 2012. "Little change in global drought over the past 60 years." *Nature* 491 (7424): 435–438.

Shimura, T., N. Mori, and H. Mase. 2015. "Future projections of extreme ocean wave climates and the relation to tropical cyclones: Ensemble experiments of MRI-AGCM3.2H." *J. Clim.* 28 (24): 9838–9856.

Sillmann, J., and M. Croci-Maspoli. 2009. "Present and future atmospheric blocking and its impact on European mean and extreme climate." *Geophys. Res. Lett.* 36 (10): L10702.

Simard, M., W. H. Romme, J. M. Griffin, and M. G. Turner. 2011. "Do mountain pine beetle outbreaks change the probability of active crown fire in lodgepole pine forests?" *Ecol. Monogr.* 81 (1): 3–24.

Spinoni, J., J. V. Vogt, G. Naumann, P. Barbosa, and A. Dosio. 2018. "Will drought events become more frequent and severe in Europe?" *Int. J. Climatol.* 38 (4): 1718–1736.

Stavros, E. N., J. T. Abatzoglou, D. McKenzie, and N. K. Larkin. 2014. "Regional projections of the likelihood of very large wildland fires under a changing climate in the contiguous Western United States." *Clim. Change* 126 (3–4): 455–468.

Stewart, I. T. 2009. "Changes in snowpack and snowmelt runoff for key mountain regions." *Hydrol. Processes* 23 (1): 78–94.

Stewart, I. T., D. R. Cayan, and M. D. Dettinger. 2004. "Changes in snowmelt runoff timing in Western North America under a 'business as usual' climate change scenario." *Clim. Change* 62 (1–3): 217–232.

Stone, D., M. Auffhammer, M. Carey, G. Hansen, C. Huggel, W. Cramer, et al. 2013. "The challenge to detect and attribute effects of climate change on human and natural systems." *Clim. Change* 121 (2): 381–395.

Strader, S. M., W. S. Ashley, T. J. Pingel, and A. J. Krmenec. 2017. "Observed and projected changes in United States tornado exposure." *Weather Clim. Soc.* 9 (2): 109–123.

Sun, F., M. L. Roderick, and G. D. Farquhar. 2018. "Rainfall statistics, stationarity, and climate change." *Proc. Natl. Acad. Sci. U.S.A.* 115 (10): 2305–2310.

Sun, L., J. Perlwitz, and M. Hoerling. 2016. "What caused the recent "Warm Arctic, Cold Continents" trend pattern in winter temperatures?" *Geophys. Res. Lett.* 43 (10): 5345–5352.

Sweet, W., G. Dusek, J. Obeysekera, and J. J. Marra. 2018. *Patterns and projections of high tide flooding along the U.S. coastline using a common impact threshold.* NOAA Technical

Rep. NOS CO-OPS 086. Silver Spring, MD: National Oceanic and Atmospheric Administration, National Ocean Service. Accessed September 30, 2018. https://tidesandcurrents.noaa.gov/publications/techrpt86_PaP_of_HTFlooding.pdf.

Sweet, W. V., R. Horton, R. E. Kopp, A. N. LeGrande, and A. Romanou. 2017. "Sea level rise." In Vol. 1 of *Climate Science Special Report: Fourth National Climate Assessment*, edited by D. J. Wuebbles, D. W. Fahey, K. A. Hibbard, D. J. Dokken, B. C. Stewart, and T. K. Maycock, 333–363. Washington, DC: US Global Change Research Program.

Sweet, W., J. Park, J. Marra, C. Zervas, and S. Gill. 2014. *Sea level rise and nuisance flood frequency changes around the United States*. NOAA Technical Rep. NOS CO-OPS 073. Silver Spring, MD: US Dept. of Commerce, National Oceanic and Atmospheric Administration, National Ocean Service. http://tidesandcurrents.noaa.gov/publications/NOAA_Technical_Report_NOS_COOPS_073.pdf.

Terando, A. J., B. Reich, K. Pacifici, J. Costanza, A. McKerrow, and J. A. Collazo. 2016. "Uncertainty quantification and propagation for projections of extremes in monthly area burned under climate change: A case study in the coastal plain of Georgia, USA." In *Natural hazard uncertainty assessment: Modeling and decision support*, edited by K. L. Riley, P. Webley, and M. Thompson, 245–256. Washington, DC: American Geophysical Union.

Trapp, R. J., N. S. Diffenbaugh, H. E. Brooks, M. E. Baldwin, E. D. Robinson, and J. S. Pal. 2007. "Changes in severe thunderstorm environment frequency during the 21st century caused by anthropogenically enhanced global radiative forcing." *Proc. Natl. Acad. Sci. U.S.A.* 104 (50): 19719–19723.

Trapp, R. J., N. S. Diffenbaugh, and A. Gluhovsky. 2009. "Transient response of severe thunderstorm forcing to elevated greenhouse gas concentrations." *Geophys. Res. Lett.* 36 (1): L01703.

Trenberth, K. E. 2011. "Changes in precipitation with climate change." *Clim. Res.* 47 (1): 123–138.

Trenberth, K. E., L. Cheng, P. Jacobs, Y. Zhang, and J. Fasullo. 2018. "Hurricane Harvey links to ocean heat content and climate change adaptation." *Earth's Future* 6 (5): 730–744.

Trenberth, K. E., A. Dai, G. van der Schrier, P. D. Jones, J. Barichivich, K. R. Briffa, et al. 2014. "Global warming and changes in drought." *Nat. Clim. Change* 4 (1): 17–22.

Trenberth, K. E., and D. P. Stepaniak. 2001. "Indices of El Niño evolution." *J. Clim.* 14 (8): 1697–1701.

USDA (United States Department of Agriculture). 2015. "*The rising cost of fire operations: effects on the Forest Service's non-fire work*." Washington, DC: USDA Forest Service. Accessed March 15, 2019. http://www.fs.fed.us/sites/default/files/2015-Fire-Budget-Report.pdf.

USGCRP (US Global Change Research Program). 2018. Vol. 2 of *Impacts, risks, and adaptation in the United States: Fourth National Climate Assessment*, edited by D. R. Reidmiller, C. W. Avery, D. R. Easterling, K. E. Kunkel, K. L. M. Lewis, T. K. Maycock, et al. Washington, DC: USGCRP.

USGS (United States Geological Survey). 2005. *Changes in streamflow timing in the western United States in recent decades*. Fact Sheet 2005-3018. Reston, VA: USGS.

Vahedifard, F., J. D. Robinson, and A. AghaKouchak. 2016. "Can protracted drought undermine the structural integrity of California's earthen levees?" *J. Geotech. Geoenviron. Eng.* 142 (6): 02516001.

van der Wiel, K., S. B. Kapnick, G. A. Vecchi, W. F. Cooke, T. L. Delworth, L. Jia, et al. 2016. "The resolution dependence of contiguous U.S. precipitation extremes in response to CO_2 forcing." *J. Clim.* 29 (22): 7991–8012.

Van Klooster, S. L., and P. J. Roebber. 2009. "Surface-based convective potential in the contiguous United States in a business-as-usual future climate." *J. Clim.* 22 (12): 3317–3330.

Van Oldenborgh, G. J., K. van der Wiel, A. Sebastian, R. Singh, J. Arrighi, F. Otto, et al. 2017. "Attribution of extreme rainfall from Hurricane Harvey." *Environ. Res. Lett.* 12 (12): 019501.

Vavrus, S. J., F. Wang, J. E. Martin, J. A. Francis, Y. Peings, and J. Cattiaux. 2017. "Changes in North American atmospheric circulation and extreme weather: Influence of Arctic amplification and Northern Hemisphere snow cover." *J. Clim.* 30 (11): 4317–4333.

Vickery, P. J., P. F. Skerjl, and L. A. Twisdale. 2000. "Simulation of hurricane risk in the U.S. using empirical track model." *J. Struct. Eng.* 126 (10): 1222–1237.

Vickery, P. J., D. Wadhera, L. A. Twisdale, Jr., and F. M. Lavelle. 2009. "U.S. hurricane wind speed risk and uncertainty." *J. Struct. Eng.* 135 (3): 301–320.

Vose, J. M., et al. 2018. "Forests." In Vol. 2 of *Impacts, risks, and adaptation, in the United States: Fourth National Climate Assessment*, edited by D. R. Reidmiller, C. W. Avery, D. R. Easterling, K. E. Kunkel, K. L. M. Lewis, T. K. Maycock, et al., 232–267. Washington, DC: US Global Change Research Program.

Vose, R. S., D. R. Easterling, K. E. Kunkel, A. N. LeGrande, and M. F. Wehner. 2017. "Temperature Changes in the United States." In Vol. 1 of *Climate Science Special Report: Fourth National Climate Assessment*, edited by D. J. Wuebbles, D. W. Fahey, K. A. Hibbard, D. J. Dokken, B. C. Stewart, and T. K. Maycock, 185–206. Washington, DC: US Global Change Research Program.

Wang, X. L., Y. Feng, G. P. Compo, V. R. Swail, F. W. Zwiers, R. J. Allan, et al. 2013. "Trends and low frequency variability of extra-tropical cyclone activity in the ensemble of twentieth century reanalysis." *Clim. Dyn.* 40 (11–12): 2775–2800.

Wang, X. L., V. R. Swail, and F. W. Zwiers. 2006. "Climatology and changes of extratropical cyclone activity: Comparison of ERA-40 with NCEP-NCAR reanalysis for 1958-2001." *J. Clim.* 19 (13): 3145–3166.

Wang, Y., B. Geerts, and C. Liu. 2018. "A 30-year convection-permitting regional climate simulation over the interior western United States. Part I: Validation." *Int. J. Climatol.* 38 (9): 3684–3704.

Watson, P. J. 2016. "Acceleration in U.S. mean sea level? A new insight using improved tools." *J. Coastal Res.* 322: 1247–1261.

Webster, P. J., G. J. Holland, J. A. Curry, and H.-R. Chang. 2005. "Changes in tropical cyclone number, duration, and intensity in a warming environment." *Science* 309 (5742): 1844–1846.

Wehner, M. F., J. R. Arnold, T. Knutson, K. E. Kunkel, and A. N. LeGrande. 2017. "Droughts, floods, and wildfires." In Vol. 1 of *Climate Science Special Report: Fourth National Climate Assessment*, edited by D. J. Wuebbles, D. W. Fahey, K. A. Hibbard, D. J. Dokken, B. C. Stewart, and T. K. Maycock, 231–256. Washington, DC: USGCRP.

Wehner, M., D. R. Easterling, J. H. Lawrimore, R. R. Heim, Jr., R. S. Vose, and B. D. Santer. 2011. "Projections of future drought in the continental United States and Mexico." *J. Hydrometeorol.* 12 (6): 1359–1377.

Westerling, A. L. R. 2016. "Increasing western US forest wildfire activity: Sensitivity to changes in the timing of spring." *Philos. Trans. R. Soc. B* 371: 20150178.

Westerling, A. L., H. G. Hidalgo, D. R. Cayan, and T. W. Swetnam. 2006. "Warming and earlier spring increase western U.S. forest wildfire activity." *Science* 313 (5789): 940–943.

Westerling, A. L., M. G. Turner, E. A. H. Smithwick, W. H. Romme, and M. G. Ryan. 2011. "Continued warming could transform greater Yellowstone fire regimes by mid-21st century." *Proc. Natl. Acad. Sci. U.S.A.* 108 (32): 13165–13170.

Westra, S., H. J. Fowler, J. P. Evans, L. V. Alexander, P. Berg, F. Johnson, et al. 2014. "Future changes to the intensity and frequency of short-duration extreme rainfall." *Rev. Geophys.* 52 (3): 522–555.

Williams, A. P., C. D. Allen, C. I. Millar, T. W. Swetnam, J. Michaelsen, C. J. Still, et al. 2010. "Forest responses to increasing aridity and warmth in the southwestern United States." *Proc. Natl. Acad. Sci. U.S.A.* 107 (50): 21289–21294.

Xu, H., N. Lin, M. Huang, and W. Lou. 2020. "Design tropical cyclone wind speed when considering climate change." *J. Struct. Eng.* 146 (5): 04020063.

Yin, J., M. E. Schlesinger, and R. J. Stouffer. 2009. "Model projections of rapid sea-level rise on the northeast coast of the United States." *Nat. Geosci.* 2 (4): 262–266.

Yonekura, E., and T. M. Hall. 2011. "A statistical model of tropical cyclone tracks in the western north pacific with ENSO-dependent cyclogenesis." *J. Appl. Meteorol. Climatol.* 50 (8): 1725–1739.

Yoshikawa, K., W. R. Bolton, V. E. Romanovsky, M. Fukuda, and L. D. Hinzman. 2003. "Impacts of wildfire on the permafrost in the boreal forests of interior Alaska." *J. Geophys. Res.* 107: 8148.

Young, A. M., P. E. Higuera, P. A. Duffy, and F. S. Hu. 2017. "Climatic thresholds shape northern high-latitude fire regimes and imply vulnerability to future climate change." *Ecography* 40 (5): 606–617.

Zhang, F., Y. Lei, Q.-R. Yu, K. Fraedrich, and H. Iwabuchi. 2017. "Causality of the drought in the southwestern United States based on observations." *J. Clim.* 30 (13): 4891–4896.

Zhou, X., and M. F. Khairoutdinov. 2017. "Changes in temperature and precipitation extremes in superparameterized CAM in response to warmer SSTs." *J. Clim.* 30 (24): 9827–9845.

Zscheischler, J., S. Westra, B. J. J. M. van den Hurk, S. I. Seneviratne, P. J. Ward, A. Pitman, et al. 2018. "Future climate risk from compound events." *Nat. Clim. Change* 8 (6): 469–477.

CHAPTER 3

Stressors and Infrastructure Resilience

One of the key messages expressed by the Fourth National Climate Assessment (USGCRP 2018) relates to the fact that damage from extreme weather events demonstrates current urban infrastructure vulnerabilities. The stresses imposed by the meteorological and hydrological hazards, described in the previous chapter, on the already aging and deteriorating infrastructure within the United States present a major challenge. This challenge is augmented when considering potential increases in the magnitudes of these stressors because of future climate change. Major stressors that will have the greatest impact on critical infrastructure include more frequent and intense precipitation events, heat waves, coastal flooding, and wildfires. The major sectors affected include our nation's energy and transportation systems, which may experience longer and more frequent power outages and service disruptions. Such impacts will have cascading effects on all other sectors, including drinking water and wastewater, transit and aviation, flood risk management, dam and reservoir operation, inland navigation, and ports and harbors. USGCRP (2018) stressed that with its long service life, urban infrastructure must be able to endure a future climate that is different from the past. Forward-looking design informs investment in reliable infrastructure that can withstand ongoing and future climate risks. Adaptation is required within each of the aforementioned sectors to prevent further critical infrastructure degradation that, in the long term, may threaten the national security, economy, essential services, and overall well-being of the United States.

Although a threshold approach to design will always be necessary in some guise, assuming that this reflects the worst-case scenario that will not be exceeded during the lifetime of the structure, it is no longer feasible or advisable. Not only are the "most probable climate conditions" difficult to define for designing future infrastructure (ASCE CACC 2015), but also other influential factors such as population growth, urbanization, and increased interdependency exacerbate the consequences of, and vulnerability to, extreme weather. The recognition that resistance is not the same as resilience has helped usher in an evolution of design to consider *safe-to-fail* (also graceful failure, see Chapter 1 and Kim et al. 2017) infrastructure or adaptive design (de Neufville and Scholtes 2011). Both concepts incorporate redundancy to facilitate reduced but continued functionality in the event of a threshold exceedance, and the latter includes scope to build future enhancements that meet evolving climate conditions (ASCE CACC 2018). This shift toward minimizing consequences from unexpected events, rather than failure avoidance (Royal Society 2014), has also been expressed across a multitude of disciplines and levels of governance (Executive Office of the President 2013, Nelson 2018, Urbano et al. 2013, World Bank 2013, WHO 2017), which all emphasize increasing global and industrial interdependence and associated vulnerability.

In assessing infrastructure vulnerability in the following, we have assumed that an ASCE grade of B−/C+ is indicative of meeting, but not exceeding, current design standards and, therefore, vulnerable to future changes. That is because although current design standards contain a "remain functioning principle" (de Bruijn et al. 2017), i.e. a minimum standard for full functionality and reduced functionality at a slightly higher target level, the reduced functionality criteria are unlikely to be met in future years. An example would be culverts that are designed to convey the 2% annual exceedance probability (AEP) (1 in 50 year equivalent) discharge but can function without failure up to the 1% AEP discharge. However, the intended life of the infrastructure is also important and flags the importance of tolerable risk. Is a 13% risk of a principal transportation artery being inundated twice in 100 years under the assumption of stationary climate conditions acceptable? When assessing future vulnerability, particularly because of changes in climate and land use and development, the changing probability and frequency of exceedance over the lifespan of critical infrastructure is even more pertinent (Volpi 2019).

The 16 infrastructure categories considered within the ASCE infrastructure report card (ASCE 2017) provided a starting point for this analysis. Because there are obvious overlaps between sectors (e.g., bridges, roads, and transit), some of these categories are combined, whereas others such as parks and recreation or schools are considered to be less critical and have not been directly addressed within this assessment. The sectors considered in this chapter were, therefore, organized into the following groups:

1. Energy transmission, storage, and distribution infrastructure;

2. Transportation—roads and bridges, and transit and aviation;

3. Drinking water and wastewater;

4. Water storage and flood protection infrastructure—levees and flood barriers, and dams and reservoirs; and

5. Navigation/ports and harbors—inland navigation and maritime navigation.

The discussion of each sector is organized into four parts: (1) potential impacts of climate change to the sector, (2) the status of sector resiliency in terms of current needs and climate change adaptation, (3) a discussion of the potential adaptation response or actions to increase sector resiliency, and (4) the dependencies and influences on other sectors that may lead to cascading failure. All the information provided for each sector can then be used along with appropriate data to conduct a prioritization or ranking of each sector; examples of prioritization schemes that have been developed and utilized in prior studies are summarized in Chapter 4.

3.1 ENERGY TRANSMISSION, STORAGE, AND DISTRIBUTION INFRASTRUCTURE

Natural disasters and climate change are a leading environmental risk to energy transmission, storage, and distribution (TS&D) infrastructure. Climate-related failures in energy TS&D infrastructure have often led to significant economic losses because of the fact that TS&D infrastructure are vital components of energy systems and other critical infrastructure nationwide. The three main sectors of TS&D infrastructure discussed in this section are as follows:

- Electric grid: The major systems of the electric grid that would be the most vulnerable to extreme weather events and climate change are electricity transmission, electricity substations, aboveground electricity distribution, and control centers.
- Natural gas: Natural gas infrastructure are, in general, less vulnerable than that of electricity grids. The majority of such infrastructure consists of underground pipelines, which are more resilient to extreme weather events. However, aboveground natural gas transmission pipelines, compressor stations, and distribution systems are key components that are vulnerable to extreme weather events and climate change.
- Liquid fuels: Liquid fuel TS&D infrastructure consists of pipelines and facilities. Regions that rely on imports will depend on the resiliency of receiving ports as well as other means of transportation.

Critical climate-related drivers impact these energy infrastructure in a variety of ways that ultimately lead to service disruptions. The critical drivers covered in this section are as follows:

- Heat waves,
- Cold waves,
- Drought,
- Extreme wind,
- Wildfires,
- Flooding,
- Sea-level rise,
- Storm surge, and
- Hurricanes.

This section discusses the potential impacts from these critical drivers, the current status of resiliency against these drivers, future initiatives in response to the critical drivers to increase resiliency related to these three sectors of the nation's energy TS&D infrastructure, and the interdependencies and influence of the energy sector on other sectors.

3.1.1 Impacts

Rising temperatures: More intense and frequent heat waves can impact energy infrastructure in a number of ways.

- Higher temperatures increase demand for air-conditioning, which, in turn, increases the load on the electric grid. This increased load reduces the efficiency of electric transmission and distribution of circuits. Extreme heat can cause transformers to malfunction.

- In thermal power plants, such as coal, nuclear, and solar plants, an increase in temperature reduces their cooling efficiency, which results in a lower power output. This increases the risk of power plant shutdowns.

- As temperatures increase, solar photovoltaic panels become less efficient. A solar photovoltaic panel at 110 °F will produce 5% to 8% less energy as compared to 77 °F (Marsh 2019). As temperatures increase because of climate change, electricity output from solar photovoltaic installations is expected to decrease.

- Pipelines in tundra regions are susceptible to the settlement of pipeline foundations and displacement of pipelines because of thawing of permafrost. This will lead to a substantial increase in operation and maintenance costs. This concern also applies to other energy TS&D infrastructure (e.g., transmission lines, fuel storage tanks, generators) in areas with permafrost.

Winter weather: Although global average temperatures are expected to increase, current trends project that winters in the United States are expected to become more variable with periods of colder and more severe weather.

- In particular, cold waves or cold snaps can cause electricity generators to fail in regions that typically do not experience freezing temperatures. Electricity generators in these regions are ill-equipped to deal with freezing temperatures, and sudden drops in temperature cause sensing lines, equipment, water lines, valves, and other components to freeze and malfunction.

- Ice storms or freezing rain is another threat to all regions of the nation. Ice storms are a significant source of power system disruptions but do not receive as much national attention as more catastrophic events such as hurricanes do. Ice storms can cause ice to accumulate on electricity transmission poles and towers. The added weight of the ice often coupled with strong winds can lead to a failure of these structures.

- Similarly, solar photovoltaic panels are susceptible to damage because of the added weight from ice storms and also to damage from hail, which can increase the costs to operate solar photovoltaic facilities.

- Extreme cold also affects natural gas production and processing. In regions that are not equipped to handle extreme cold, freezing temperatures cause residual water in gas to freeze, restricting the flow of gas at production wells.

- In addition, natural gas processing plants are susceptible to mechanical failures in extreme cold.

Precipitation and drought: The distribution of precipitation across the country is expected to become more variable because of climate change, resulting in increased flooding in some areas and drought in others.

- Electricity substations in flood-prone areas will experience more disruptions as the intensity and frequency of flooding increase.

- Aboveground energy product storage will also experience increased vulnerability because they can be moved off their foundations, resulting in spills and leaks.

- Extreme rainfall events and increased melting of snowpack will lead to flooding in many areas. Because hydroelectric facilities rely on water for power generation, they are particularly susceptible to extreme rainfall events and flooding. During times of increased rainfall, water may need to be diverted into spillways to prevent overtopping of dams (EPA 2017). Increased water levels will place higher demands on both dams and spillways.

- More severe and frequent droughts will directly affect hydroelectric generation by decreasing the volume and pressure of water available to drive turbines, thus decreasing power output. This is particularly true in the northwest and southwest regions of the country where hydroelectric generation relies heavily on mountain snowpack, which is expected to decrease (USGCRP 2018)

- Natural gas processing via fracking also relies on water availability, and more frequent shortages of water will limit natural gas processing.

- Droughts can also impede the transport of energy products that rely on barges.

Wind: Extreme wind events are also expected to increase in severity and intensity. Extreme wind events are also associated with other types of climate events. Hurricanes, tropical storms, and tornadoes are expected to increase in severity because of climate change. Increased temperatures create conditions for stronger winds during hurricanes and tropical storms because of increased ocean surface temperatures and stronger tornadoes because of increased instability in the lower levels of the atmosphere (Biello 2013).

- Increase of extreme wind events primarily from hurricanes, tropical storms, and tornadoes will, in turn, increase the risk of damage to many structures critical to the TS&D infrastructure, including power plants, liquid fuel storage tanks, pipeline systems, and electric grids. Electric poles and transmission lines are the most vulnerable components of an electric grid, and an increase in damaging wind events will also increase blackouts.

- Climate change is expected to alter current wind patterns, complicating the installation of wind farms across the United States. The construction of wind farms will need to rely on climate forecasts that account for the changing wind patterns that will affect the location, distribution, and sizing of wind farms and turbines, among other factors. An increased number of high wind events will lead to the potential for damage to wind turbines. In addition, by 2100, potential wind energy in the United States is expected to fall by approximately 10% to 40% because of a decrease in the temperature variation between the North Pole and the equator (Karnauskas et al. 2018).

- Because of the relatively large surface areas of solar photovoltaic panels, they are subjected to high wind loads with the potential for damage. Their high surface area-to-weight ratio makes them susceptible to wind-induced vibrations, which can lead to reduced energy output and damage to the panels and supporting structures.

Hurricanes: Current predictions estimate that hurricanes will become more frequent and intense because of climate change. This trend, coupled with sea-level rise, will lead to increased flooding and storm surges.

- Increase in hurricane intensity increases coastal electricity substation exposure to storm surge and wind damage. Exposures of electricity substations to storm surge are also expected to increase because of sea-level rise.

- Offshore oil and gas platforms are highly vulnerable to the effects of hurricanes, and increases in the intensity of hurricanes will only exacerbate these effects. As a result, oil and gas platforms may experience more significant downtime during and after hurricanes.

- Offshore wind farms are a growing area of interest primarily because of the potential for higher and more constant wind speeds, leading to highly reliable energy output. However, the increased intensity of hurricanes will lead to higher risks of failure because of both wind and wave action.

Wildfires: Wildfires will become more severe because of the projected increase of high wind events and droughts.

- Increase in the intensity and frequency of wildfires poses an increased risk of damage to electricity transmission systems such as poles, transmission lines, transformers, and substations.

- Wildfires are also a threat to aboveground natural gas measurement and control instrumentation, and damage can lead to natural gas shutoff. The western region of the United States is particularly vulnerable to electricity and gas disruptions because of wildfires. Wildfires can lead to the destruction of gas and electricity infrastructure, but electricity distribution lines can also lead to wildfires, as was the case in the California Camp Fire Wildfire in November of 2018 in which investigators found that the fire was caused by Pacific Gas and Electric (PG&E) powerline failures during high winds. Not only do wildfires lead to power outages, but PG&E also resorts to elective power outages during periods of high winds and dry conditions as a precautionary measure against wildfires. About 60,000 people were left without power during one such power outage in 2018 (Fimrite and Palomino 2018). These types of power outages are expected to increase in the future because of climate change.

3.1.2 Status

At present, resiliency efforts for electric grids are mainly centered on the lower end of the technology spectrum. This includes building physical protective barriers, implementing more subterranean lines, strengthening or reinforcing poles, promoting aggressive vegetation management, and investing in backup systems.

Natural gas infrastructure currently use sensors and automated response systems as a resilience measure in the event of disruptions. These seemingly simple components of natural gas infrastructure are effective in shutting off disrupted areas, and as a result, gas can be redistributed relatively easily. Inherently, natural gas infrastructure is less susceptible to disruptions because the system has very few single points of failure that would impact large portions of entire systems. Therefore, disruptions can be isolated relatively easily, with minimal impact to end customers in most cases. Also, natural gas wells and storage facilities are widespread, and gas lines can maintain high pressure even in the event of failure of half of the compressors, making them highly redundant. Another effective natural gas resiliency measure is the usage of backup gas-fired combined heat and power (CHP) generators. CHP generators performed quite well during the recent hurricanes Harvey and Irma. These generators were instrumental in supporting critical facilities such as hospitals immediately following the hurricanes (NGC 2017).

For liquid fuels, there is a range of actions that are currently employed to address the vulnerabilities outlined in the previous section. Stockpiling liquid fuel products is a relatively easy strategy that requires little investment in new technologies. For example, the United States maintains a national emergency fuel storage reserve along with other regional reserves. Many countries besides the United States have also adopted this strategy. However, there is currently much debate among experts regarding how large the stockpile of liquid fuels should be. Other hardening strategies such as additional flood protection, backup electric power generators, and so on have also been commonly implemented.

3.1.3 Response

New initiatives to improve resiliency to the electric grid focus on new technologies and improved modeling and data analysis. Advancements in smart meters, automated switches, microgrids, and related technologies are a few examples of technologies that can significantly increase electric grid efficiency and resiliency. The electric grid is a complicated network of devices linked together by control systems. The aforementioned technological advancements reduce inherent inefficiencies, which are even more critical during disruptions. Advancements in modeling and data analysis are critical to support advancements in system operation and control and will provide the electric grid with a better ability to automatically respond in real time to extreme weather events. Modeling and data analysis initiatives seek to improve the visibility of real-time data both temporally and spatially across the grid. Part of these initiatives also includes improving and optimizing current climate forecasting methods. Improved forecasting methods would operate in concert with smart technologies to further reduce inefficiencies. Improvements in microgrid technologies and methods will allow electricity disruptions to be isolated to smaller areas in contrast to widespread blackouts and help to minimize economic losses. Critical to the success of microgrids is the ongoing research to develop more efficiently distributed energy resources (DERs). Renewable DERs are a major focus of a lot of the ongoing developments. DERs are sources of energy that are independent of the main electric grid. More efficient DERs will better supply more sections of the electric grid and allow them to operate independently of the main grid during a disruption.

Similar to the electric grid, research and improvements to natural gas operation and control technologies are on the horizon (e.g., smart meters, smart sensors, automated switches). Improvements in these technologies would similarly allow the natural gas infrastructure to be more efficient during disruptions by redistributing gas supply quicker and more efficiently. Improvements and more widespread implementation of smart sensors, satellite imaging, and aerial drones will also enable repair crews to identify gas leaks faster, reducing natural gas downtimes.

The majority of the liquid fuel TS&D infrastructure initiatives aim to analyze the economic impacts of emergency reserves. Although ongoing cost–benefit analysis may seem trivial and redundant, these analyses are constantly including more up-to-date research regarding climate change and improvements in liquid fuel infrastructure to make the most up-to-date recommendations. Just within the last 20 years, the United States has created two new regional petroleum product reserves as a result of cost–benefit analysis (DOE 2015).

In some regions, the energy sector has taken proactive measures to harden and increase the resiliency of the energy infrastructure for the effects of climate change. Hardening in this context is defined as physically changing the infrastructure to make it less susceptible to damage from extreme weather (Wang et al. 2016). Examples include constructing physical barriers to protect equipment and generation facilities from storm surges, elevating and installing waterproof equipment in areas of potential flooding, reinforcing equipment to protect against increased wind loads, burying power lines underground, and constructing more distributed energy storage facilities. Resiliency is defined as the ability of an energy facility to recover quickly from damage. Examples of resiliency measures include planning and training to deal with failures, evaluating existing infrastructure to identify critical components, managing vegetation, conducting inspections, and procuring emergency backup equipment (Wang et al. 2016). The future outlook will rely heavily on designers and the construction industry to develop, design, and construct new and innovative solutions.

New York City (NYC) provides one example of the types of innovative measures being implemented to prepare for climate change. NYC has released a set of climate resiliency design guidelines, which provides predictions of changing weather patterns and design recommendations to prepare for climate change (ORR 2019). The design recommendations include the following:

- Higher temperatures will increase the demand for air-conditioning, which will likely lead to increased blackouts and brownouts. Buildings and infrastructure should be designed to accommodate periods without electricity. Critical load items and install backup power systems should be included to supply the necessary power.

- Passive solar cooling and ventilation in the design of buildings should be included to both reduce energy demands and better cope with periods of no electricity.

- Increased flooding is expected from more extreme precipitation, hurricane storm surges, and rising sea levels, which can flood and destroy underground and grade-level utilities. Underground and grade-level utility infrastructure should be designed to be impervious or resistant to water damage including saltwater. Additional measures include elevating utilities above predicted storm surge elevations by installing them on the upper floors of a building and relocating utilities outside of floodplains, which are expected to expand.

The combination of a growing population, reduced efficiency of thermal electric power generation and transmission, and increased electricity demand because of higher temperatures and the transition to electric vehicles will result in the need for additional power generation facilities. By some estimates, construction of up to 25% more capacity may be required by 2040 (Larsen et al. 2017). This new electricity generation capacity will further exacerbate the impacts of climate change unless it is supplied via carbon-free sources.

Fortunately, the costs of electricity generation via renewable sources such as wind and solar have dramatically declined in recent years because of advancements in technology and increased production and the associated economies of scale. By some estimates, the costs of energy production from utility-scale solar plants declined by 86% from 2009 to 2017 (Lazard 2017). One study found that, in 2018, electricity produced by wind or solar was cheaper than that produced by coal in 74% of cases throughout the United States (Gimon et al. 2019). In 2016, electricity generation from solar resources grew by 44% and that from wind resources grew by 19% (EIA 2017). Overall, in 2018, electricity generation from renewables in the United States grew to its highest level of 18%, which, in 2016, was up from 15%. Solar resources accounted for 60% of that capacity, and wind resources accounted for 38% (Zindler 2019). Eight states including Vermont, Washington, California, and Oregon are well beyond the national average, obtaining approximately 50% or more of their electricity from renewable resources (EIA 2018). In recent years, many states have passed laws pledging to increase electricity production from renewables. By 2045, Hawaii has pledged to become carbon-neutral.

Increased focus and development of renewable resources other than wind and solar resources is also being witnessed, with examples being energy generation from waves and tides as well as geothermal resources. Energy production from a variety of renewable resources will not only help to lower carbon emissions but also make the energy infrastructure more resilient by providing more reliable sources of energy.

3.1.4 Interdependencies

The energy infrastructure is characterized as uniquely critical because of the dependence of all other critical infrastructure on its functionality (illustrated in Figure 3-1), and the resilience of the energy infrastructure in the face of extreme weather and climate change is of utmost importance. As demands on the energy infrastructure increase because of climate change, the reliability of the energy sector may be compromised if actions are not taken to improve its resiliency. This reduced reliability will have downstream effects on all sectors that rely on electricity and liquid fuels (e.g., transportation, water, wastewater, manufacturing). Those sectors that are directly powered by electricity will be immediately affected, whereas those relying on liquid fuels will be capable of sustaining normal operations only for a limited period of time depending on storage reserves.

The energy sector is also reliant on many of these sectors for its operation. For example, this sector relies on the transportation sector (highlighted in Figure 3-1 as a primary interdependency) for the transportation of energy products as well as operation and maintenance of existing energy infrastructure. The transportation sector will be affected by many of the same forces directly impacting the energy sector such as extreme wind, flooding, drought, wildfires, and so on. Energy

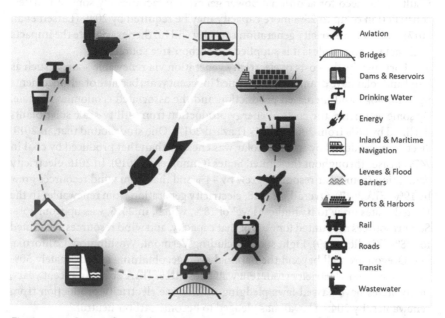

Figure 3-1. Sectoral interdependencies involving the energy sector. The arrow thickness indicates the level of criticality of sector interdependence. The double arrow indicates the dependence of both sectors on each other.

products are transported mainly through pipelines, barges, rail, and trucks, and a disruption in the infrastructure supporting these four means of transportation will also lead to a disruption in the transmission of energy. Repairing damaged energy infrastructure, especially because of extreme weather events, is often highly time-sensitive and relies heavily on the proper functioning of the transportation infrastructure.

Another key co-interdependency is with the drinking water and wastewater sectors (highlighted as a secondary interdependency in Figure 3-1). These sectors are critical because the short-term impacts of any disruption within these sectors can have major consequences. Therefore, ensuring continuous operation and functionality within the drinking water and wastewater sectors is of utmost importance and depends foremost on the supply of electricity.

Other interdependencies identified in Figure 3-1 include the dependence of the energy sector on flood protection and the dependence of the transit and aviation, ports and harbors, and inland and maritime navigation sectors on an uninterrupted source of electricity and energy products for fuel. For example, major flood events can result in widespread power outages that will have major impacts on all other sectors; the existence of flood protection infrastructure near power stations located in flood-prone areas is key to preventing such events. Maintaining an uninterrupted operation of dams and reservoirs is also important in those areas that are dependent on hydropower for a large percentage of their energy needs. The energy sector also depends on water supply for cooling purposes and, hence, a double arrow is indicated between energy and dams and reservoirs.

3.2 TRANSPORTATION

3.2.1 Roads and Bridges

Although changes in mean climate will have some impacts on the transportation (specifically roads and bridges) sector, for example, through more rapid degradation of materials, a greater risk is posed by extreme events. The infrastructure report card (ASCE 2017) focuses on the current status of the roads from a day-to-day user perspective—for example, wearing surface, congestion, and fatalities. All these will likely deteriorate with changes in the climatology (i.e., the weather expected over a 30 year period), variability, and extreme weather events. However, the infrastructure is also vulnerable to the immediate effects of extreme weather events, such as floods and consequent landslips or subsidence. A thorough examination of the observed and likely impacts of extreme weather on the transportation network and estimated associated costs is presented in Chapter 12 of the National Climate Assessment (Jacobs et al. 2018a).

Nuisance flooding has also become a matter of increasing concern, both in coastal and in inland regions (Moftakhari et al. 2017, 2018). Although minor disruptions to the transport system may be a source of annoyance only in urban areas, rural communities may suffer disproportionately because of the lack of alternatives (Mattsson and Jenelius 2015). The additional operational and maintenance burden presented by frequent weather hazards is also more easily absorbed within urban areas, where the higher density of population also offers a higher-density tax resource. However, smaller rural and local authorities under increasing budgetary pressure find that they have to juggle the needs of maintaining principal routes open and tackling the problem of degradation, and, hence, reduced resilience, of less populated routes. Although the cumulative impacts of nuisance events can be as high, or higher, as those from rarer weather events, there is no mechanism to account for these impacts in a traditional cost–benefit analysis. Furthermore, the long-term impacts on network performance, for example, arising from enhanced loads on diversion routes, are poorly understood.

Impacts on the road and bridge network can be classified into the following, where each impact can be considered from a failure, maintenance, or usability perspective (Bles et al. 2016):

- Embankments and foundations,
- Road and bridge surface course/pavement, and
- Peripheral supporting systems.

Drivers that are expected to have the greatest impact on roads and bridges are the following :

- Extreme precipitation (solid, liquid, and drought),
- Temperature extremes,
- Freeze–thaw cycles,
- Storm surges, and
- Extreme wind speeds.

3.2.1.1 Impacts

Increases in the frequency and spatial extent of highly intense precipitation (Du et al. 2019, Pendergrass and Knutti 2018, Prein et al. 2020), interspersed with longer dry periods, pose numerous threats to road and bridge infrastructure (Table 3-1). At the shortest time scale (minutes to hours), heavy precipitation can lead to reduced skid resistance, aquaplaning, and poor visibility, which can all reduce traffic safety and lead to a higher frequency of collisions. Such storms can also overwhelm drainage systems, leading to localized flooding and possible pollution, and affect the stability of earthworks that are currently under construction or of recently planted shoulder vegetation.

Repeated flooding, from overwhelmed surface water drainage, riverine, or coastal sources, can lead to a loss of structural integrity, bridge pier scouring, a weakening of embankments and foundations, or an uplift of tunnels or lightweight construction materials. In contrast, rarer precipitation or flood events can lead to immediate embankment failure, can wash away sections of highways or bridges, can lead to landslips and rock falls, and cut access to such places and

Table 3-1.   *Climate Variables Posing Risks to Road and Bridge Infrastructure.*

Critical climate variables	Risks to road and bridge infrastructure
Extreme rainfall events (heavy showers and long periods of rain)	Flooding of roadways Road erosion, landslides, and mudslides that damage road Overloading of drainage systems, causing erosion and flooding Traffic hindrance and safety
Seasonal and annual average rainfall	Impact on soil moisture levels, affecting the structural integrity of roads, bridges, and tunnels Adverse impact of standing water on the road base Risk of floods from runoff, landslides, slope failures, and damage to roads if changes occur in the precipitation pattern (e.g., changes from snow to rain in winter and spring thaws)
Sea-level rise	Inundation of roads in coastal areas Erosion of the road base and bridge supports Bridge scour Reduced clearance under bridges Extra demands on the infrastructure when used as emergency/evacuation roads
Maximum temperature and number of consecutive hot days (heat waves)	Concerns regarding pavement integrity, for example, softening, traffic-related rutting, embrittlement (cracking), and migration of liquid asphalt Thermal expansion in bridge expansion joints and paved surfaces Impact on landscaping
Drought (consecutive dry days)	Susceptibility to wildfires that threaten the transportation infrastructure directly Susceptibility to mudslides in areas deforested by wildfires Consolidation of the substructure with (unequal) settlement as a consequence More generation of smog Unavailability of water for compaction work
Snowfall	Traffic hindrance and safety Snow removal costs Snow avalanches resulting in road closure or striking vehicles Flooding from snowmelt

(Continued)

Table 3-1. Climate Variables Posing Risks to Road and Bridge Infrastructure. (Continued)

Critical climate variables	Risks to road and bridge infrastructure
Frost (number of icy days)	Traffic hindrance and safety
	Ice removal costs
Thaw (number of days with temperature zero crossings)	Thawing of permafrost, causing subsidence of roads and bridge supports (cave-in)
	Decreased utility of unimproved roads that rely on frozen ground for passage
Extreme wind speed (worst gales)	Threat to the stability of bridge decks
	Damage to signs, lighting fixtures, and supports
Fog	Traffic hindrance and safety
	More generation of smog

Source: Bles et al. (2010).

terrain. Heavy precipitation falling on postwildfire hydrophobic soils can also lead to localized flooding and debris flows, blocking or damaging culverts and greatly exacerbating the impacts of subsequent precipitation. Given both observed and projected increases in the frequency and intensity of precipitation events, coupled with urbanization (and, hence, reduced soil porosity), or increased risks of drought and wildfires, the probability of current design criteria being exceeded will also increase (Lopez-Cantu and Samaras 2018).

Increasing winter temperatures in the rain–snow transition zone will also increase the frequency of heavier, wetter snow in addition to the frequency of rain-on-snow events. In addition to increasing the volume of snow water equivalent, and, hence, the potential meltwater flood volume, the former will increase loading on bridges and peripheral structures, potentially increasing structural failures, and may exacerbate the winter maintenance burden. Increased frequency of rain-on-snow events will exacerbate some of the worst flooding and associated impacts listed in previous paragraphs in regions such as the midwest—as witnessed in winter 2016 in the Missouri and Mississippi river basins.

Sea-level rise is already affecting many coastal areas with *sunny day flooding*, exacerbating congestion and damage to infrastructure, in addition to preventing access to neighborhoods or through tunnels during flood events (Behr et al. 2016, Jacobs et al. 2018a). It is estimated that by 2060 7,500 mi of highway currently affected on the US East Coast will increase by a factor of 10 (Jacobs et al. 2018b). These impacts, combined with increases in storm surge, from tropical cyclones or coastal storms, also contribute to the inundation of road surfaces, erosion of embankments, and bridge scour. In the most northern parts of Alaska, the combined effects of sea-level rise and longer sea-ice free seasons are contributing to rapid increases in coastal erosion and subsequent damage to highway infrastructure, particularly from severe winter storms (Gibbs and Richmond 2016, IAWG 2009, USACE 2009).

Increasing temperatures bring a number of risks at both extremes of the temperature spectrum: from increases in flood risk because of rapid snowmelt, reduced exposure, and, therefore, the ability to manage cold extremes to extended periods of very high temperatures, compromising health and safety and accelerating structural degradation. Higher ambient temperatures in the southern United States are already affecting pavement performance, concrete integrity, and, in combination with changes in salinity and humidity, the structural condition of bridges. With regard to driving, increased temperatures can cause the tires of vehicles to shred and vehicles to overheat, whereas rutting and misalignment of expansion joints can make driving surfaces less safe. The costs of pavement maintenance and repairs following the 2011 heat wave in Texas were estimated at $26 million for the Texas Department of Transport (DOT) (NASEM 2014); this does not account for the other incurred costs of delays, injuries, or vehicle damage.

In some regions, notably in the more southern regions, increases in winter ambient temperatures will bring modest benefits in the form of a reduced frequency of frost days and associated maintenance and of perilous snow driving conditions. However, as demonstrated by the cold winter snaps of 2013/2014 and 2017/2018, freezing and snow conditions will still occur albeit with less frequency. This reduction in exposure is likely to lead to less preparation on the part of DOTs and less ability by the general public to cope with extremely cold conditions. In contrast, more northerly regions will face greater winter maintenance challenges where they have previously not experienced freezing snow and rain, or substantial freeze–thaw damage. The effects of temperature are also already being experienced in Alaska, where melting permafrost is causing considerable road surface and pipeline damage. Similarly, in the Midwest, Northeast, and Northern Great Plains, the reduction in the frequency of extremely cold days is affecting transportation through a reduction in allowable winter vehicular weights on frozen roads and ice roads (i.e., semipermanent roads on a frozen water surface).

Along the coast and proximate to large bodies of water (e.g., the Great Lakes), changes in extreme winds will affect storm surge and wave action, with the consequences previously listed Increases in the frequency of wind storms or their intensity will also have direct impacts on road and bridge infrastructure and their users, for instance, through the number of bridge closures to high-sided vehicles, potentially increasing freight costs. Increases in wind storm frequency, in combination with increased precipitation and ground saturation, will also increase the probability of damage from falling trees, pylons, or other isolated structures. Effects will range from reduced access while the fallen item is removed to serious damage of structural components (e.g., bridges). High winds also pose a threat to sign posts and lighting infrastructure, potentially increasing the maintenance budget, as well as posing a threat to the driving public from airborne materials.

3.2.1.2 Status

The transportation sector is one of the most advanced sectors in terms of planning for and assessing the vulnerability of its assets (Jacobs et al. 2018a). Major cities have potentially led the charge in this regard given the high densities of population and dependence on transportation infrastructure. Rural communities, with higher budgetary constraints, are less well developed in their assessments of current vulnerabilities. The degree of completion and thoroughness of vulnerability assessments across the country varied. States with a higher density of population, in closer proximity to coastlines (e.g., Maryland; Washington, DC; Florida; and California), or that have experienced more recent catastrophic events (e.g., Colorado) tend to be more advanced in their assessments. However, very few have moved beyond vulnerability assessments to develop probable adaptation measures.

Federal Highways have partnered with several State DOTs and Metropolitan Planning Authorities on 11 pilot projects to assess vulnerabilities, examine resilience measures, and eventually implement these measures. The program for implementation and monitoring will run from 2020 to 2024 (FHWA 2018). Although the 11 pilot projects are focused on resilience and durability to extreme weather, the main areas of focus are flooding from different sources (fluvial, pluvial, and coastal) and wildfire. Some impacts from changes in temperature will be examined, but an increased knowledge of the likely responses to changes in wind patterns and snow loading, or potential solutions, will require further examination in later projects.

The National Flood Insurance Program (NFIP) maps provide the most commonly used information on probable coastal and fluvial flood risk. However, in many cases, the data are out of date or do not account for likely increases in exposure with sea-level rise. Furthermore, they do not account for all sources of flooding—such as surface water (pluvial) or known sewerage overflows—and were developed for insurance underwriting rather than for assessing engineering risks (ASCE CACC 2018). That is, the maps only indicate an estimated envelope of land below a target water surface elevation and do not include information on flow direction, velocity, depth of flooding, or surge loading, all of which are crucial for structural damage and public safety analyses.

Because of the individual component approach to risk analysis that has been commonly adopted in the past, discrepancies often arise affecting network resiliency. For instance, bridge infrastructure is commonly designed to convey the 1% annual probability flood flow, without impacting a bridge's structure. However, the approach roads to bridges are not always elevated to the same degree above the floodplain.

3.2.1.3 Response

Several recent reports have addressed possible adaptation measures for highways and bridges (NYSDOT 2011, Bles et al. 2010). In addition to policy, design guide, and management or maintenance changes, some specific structural adaptations can be explored, as summarized in Table 3-2.

Table 3-2. Summary of Impacts to Road and Bridge Infrastructure from Climate Change and Potential Remedial Actions.

Infrastructure impact	Operations impact	Potential response
Increasing frequency and/or intensity of extreme precipitation		
Increased rate of deterioration of materials (e.g., road base) or increased structural degradation	Increased maintenance cycle frequency and associated costs	Change in materials selection; improved drainage capacity; advances in the use of green infrastructure for contaminant control; increased inspections and maintenance
Damage to roads, tunnels, and bridges, or obstruction from floods, increased erosion of road and bridge embankments and foundations, landslides, and slope failures	Increasingly frequent disrupted access; traffic delays; loss of access routes; increased maintenance/repair costs	Increase the capacity of drainage structures; advances in the use of green infrastructure for flood storage and management; strengthening side slopes; increased inspections and maintenance; changes in design standards to accommodate higher flood flows and/or ensure redundancy; raising or relocating bridges and highway structures away from floodplains; redesigning bridge scour protection systems
Decrease in skid resistance, reduced vehicle control	Increasing frequency of poor visibility from precipitation or splash and spray; aquaplaning because of standing water on road surfaces	Change in materials selection; increased use of porous/permeable pavement; improved drainage capacity; changes in design standards to accommodate higher precipitation volumes

(Continued)

Table 3-2. *Summary of Impacts to Road and Bridge Infrastructure from Climate Change and Potential Remedial Actions. (Continued)*

Infrastructure impact	Operations impact	Potential response
Increased frequency or duration of temperature extremes		
Loss of road structure integrity	Increased maintenance for unequal settlement of roads and foundations because of drought (consecutive dry days); increasing frequency of closures and damage from wildfire	Changes in material selection; reviewing the design life, foundation design, and surface course design of flexible pavements
Loss of pavement integrity through thermal expansion, cracking, rutting, or embrittlement	Increasing frequency of damage or thermal expansion because of higher maximum and minimum diurnal temperatures	Increased thermal expansion joints in concrete pavements and bridges; changes in material selection
Decrease in skid resistance	Increased frequency of migration of liquid bitumen in high temperatures	Changes in material selection
Reduced ability for maintenance and construction	Increased maximum and minimum diurnal temperatures; increased frequency of consecutive hot days (heat waves)	Monitoring construction and maintenance projects to reduce heat exposure
Increased winter minimum temperatures and changes in snow patterns		
Loss of pavement integrity	Increased cracking from frost–thaw cycles; increased detachment of pavement layers, and frost heave	Increased design snow loads; changes in material selection

(Continued)

Table 3-2. Summary of Impacts to Road and Bridge Infrastructure from Climate Change and Potential Remedial Actions. (Continued)

Infrastructure impact	Operations impact	Potential response
Increased rate of deterioration of materials (e.g., road base) or increased structural degradation	Increased frequency of frost–thaw cycles reducing winter loading capacity; increasingly wet snow or consecutive events	Increased design snow loads; increased foundation depth; a review of winter loading criteria; changes in material selection
Loss of driving ability	Increased frequency of snowstorms (in northern and midwest states) reducing visibility; decrease in skid resistance; reduced access from snow-blocked roads	Reallocating resources for snow and ice operations; installing intelligent signs or other information systems; improving coordination with weather forecasting
Sea-level rise and storm surge		
Increased rate of deterioration of materials (e.g., road base) or increased structural degradation with saline incursion	Increased maintenance of cycle frequency and associated costs	Change in materials selection; improving drainage capacity; increasing inspections and maintenance; raising or relocating bridges and highway structures away from floodplains; constructing local levees, sea walls, or floodgates

(Continued)

Table 3-2. Summary of Impacts to Road and Bridge Infrastructure from Climate Change and Potential Remedial Actions. (Continued)

Infrastructure impact	Operations impact	Potential response
Damage to roads, tunnels, and bridges, or obstruction from floods, increased erosion of road and bridge embankments and foundations, landslides and slope failures; uplift of tunnels or lightweight construction materials	Increasingly frequent disrupted access; traffic delays; loss of access routes; increased maintenance/repair costs	Increasing the capacity of drainage structures; strengthening side slopes; increasing inspections and maintenance; raising or relocating bridges and highway structures away from floodplains; constructing local levees, sea walls, or floodgates; installing bridge impact-protection systems; redesigning bridge scour protection systems; reviewing design criteria for buoyancy changes
Extreme wind speeds		
Damage to ancillary structures (e.g., signs and lighting fixtures)	Increased frequency of high wind gusts	Increasing design wind loads; reviewing foundation designs
Loss of driving ability; reduced vehicle control (especially high-sided vehicles)	Increased frequency of high wind gusts or sustained high wind speeds	Installing anemometers and smart monitoring; reviewing wind velocity limits and restrictions
Damage to roads or bridges, or obstruction from fallen trees and other large objects	Increased frequency of high wind gusts in combination with soft ground layers	Increasing design wind loads; reviewing foundation designs

3.2.2 Transit and Aviation

Land- and air-based modes of public transportation covered in this section include light rail, heavy rail, bus transport, and commercial airline services. Brief descriptions of these modes of public transportation are as follows:

- Light rail: electric traction streetcars, trolleys, and tramways in which cars operate singly (or in short, multicar trains) on fixed rails in the right of way and often separated from other traffic for a part or much of the way (FTA 2017).

- Heavy rail: electric multicar trains with the capacity to handle a heavy volume of traffic and characterized by exclusive rights of way, high speed, rapid acceleration, sophisticated signaling, and high platform loading, termed as *commuter rail* and can be underground, at-grade, or elevated (FTA 2017).

- Bus transport: self-propelled, manually steered vehicles in shared rights of way with fuel supply carried on board (APTA 1994).

- Commercial airline services: scheduled passenger and cargo flights operating between commercial airports within controlled airspace.

The functioning of these services relies upon various subsets of *vertical* (buildings and facilities) and *horizontal* (roads, track, bridges, energy transmission) infrastructure with the potential to be directly impacted by climate change stressors and indirectly impacted through interconnection with other infrastructure categories. The following climate change stressors have the potential to significantly impact transit and aviation infrastructure:

- Increasing heat because of climate warming and urban heat island effects,

- Increasing temperature volatility and changing freeze–thaw cycles,

- Increasing intensity and frequency of precipitation events,

- Increasing frequency of tidal flooding because of sea-level rise, and

- Increasing intensity and frequency of extreme coastal flooding events.

This section presents the potential impacts of each of these climate change drivers on transit and aviation, the status of America's navigation infrastructure regarding resilience against these drivers, and potential future responses to increase resiliency.

3.2.2.1 Impacts

The impacts of these climate change include extreme events (e.g., coastal flooding, rainfall, heat, and wildfires, which are projected to increase in frequency and/ or magnitude) that would exceed transit and aviation infrastructure design thresholds (e.g., result in a reduction in infrastructure performance or catastrophic failure), as well as longer-term impacts because of gradually increasing and persistent stressors (e.g., as increasing average temperature, temperature volatility, more frequent rainfall flooding, and/or higher tides because of sea-level rise) that have the potential to impact infrastructure maintenance and useful life. These impacts can be viewed in terms of the physical effects on the transit and aviation infrastructure itself, as well as the effects on resultant operations because of reduced infrastructure performance or failure.

Increased heat will have a variety of impacts that may require changes in design, equipment, materials, structural design, maintenance, and/or operations. The effects of an increase in mean temperatures, as well as the frequency of extremely high temperatures, have the potential to increase the rate of deterioration of certain materials (e.g., concrete and asphalt); cause exposed rails to crack, buckle, pull apart, or separate; cause overheating of building and vehicle HVAC systems, electrical equipment such as substations, switchgears, and signals, failure embankments, and shallow foundations (in permafrost areas); and increase the risks of aviation incidents because of increased wind gusts and less dense (hotter) air that challenge airplanes' ability to generate lift.

Increased temperature volatility will challenge the performance of materials that are susceptible to thermal expansion and contraction, resulting in increased wear and tear of the materials themselves and the associated connected systems (e.g., tracks, building facades, concrete slabs, and pipelines). More frequent and intense freeze–thaw cycles have the potential to increase heaving or rutting of pavement. In addition, increased swings from low temperature to high temperature following precipitation can increase the occurrence of ice formation on roads, tracks, airport runways, and supporting infrastructure, leading to the need for increased maintenance, health, safety, and service disruptions.

Increased intensity and frequency of precipitation and tidal flooding because of sea-level rise has the potential to result in more frequent and severe flooding of low-lying airports, roads, rail, and associated infrastructure (e.g., signals and substations), particularly if they operate in concert with one another. Both stressors (most notably sea-level rise for tidally influenced locations) also have the potential to cause seepage flooding or ponding because of a rising groundwater table. Tunnels and underground transit systems are particularly vulnerable to flooding, resulting in threats to rider and worker health and safety, damage, and service disruptions because of inaccessibility and malfunctions of systems including signaling and sensing equipment.

Increased intensity and frequency of extreme coastal flooding has the potential to result in catastrophic flooding of assets in coastal areas that are not sufficiently elevated. Storm surges can result in very high overland and subgrade inputs of saltwater, introducing significant hydrostatic, hydrodynamic, breaking wave, impact, and corrosive loads on infrastructure. Major flooding of infrastructure storage and maintenance facilities; substations; tunnels and underground transit systems, including ventilation buildings; and airport terminals, runways, emergency power installations, and interterminal transit networks threaten life and force operations to shut down for long periods of time until the infrastructure can be rebuilt.

3.2.2.2 Status

By nature, land-based transportation infrastructure relies on interconnected systems to facilitate the transfer of energy, people, and materials across a variety of spatial scales. Transit buses, metros, and subways within urban centers operate on a relatively small geographic scale but have a high volume. Service disruptions at this level can have large and cascading impacts because of these high flow densities, combined with interconnected modes of transport. However, an urban transportation network can have increased redundancy because of the likely availability of multiple transit options (including, in some cases, walking). Passenger railroad service and coach buses that provide medium- and long-distance intercity service have broader spatial distribution but have fewer passengers per day and larger headings between cars. The broader distribution of infrastructure can present broader vulnerabilities to climate change drivers and maintenance challenges. In addition, rural communities may rely more heavily on transportation options because of a lack of redundancy.

A relatively high percentage of the transit infrastructure is aligned within valleys, through tunnels and/or land cut to lower elevations than the surrounding areas, which is often characterized by limited drainage and susceptibility to flooding. Likewise, many airports and much of the supporting infrastructure for transit systems (e.g., train/bus maintenance facilities and storage yards) have been constructed within historically flat low-lying areas (e.g., filled wetlands and floodplains) for practical and economic reasons. Airports require expansive flat land for runways to provide room for aircraft to take off and land and are typically situated near urban centers (most major cities are close to the sea or large rivers). These factors leave a high percentage of transit and aviation infrastructure vulnerable to precipitation, sea-level rise, and extreme coastal flooding events.

3.2.2.3 Response

Table 3-3 provides the impacts and potential responses of climate change on transit and aviation identified in the literature.

3.2.3 Interdependencies

In addition to energy, highways (and associated structures) are arguably one of the most important sectors with regard to interdependency. Although they may rely heavily on energy infrastructure to maintain power supply for supervisory control and data acquisition controls, maintenance facilities, or moving bridge operations, for example, the energy sector, in turn, depends on the transportation of energy raw materials and products as well as road access to allow continuous operation and maintenance of existing energy infrastructure; this codependence is illustrated with the heaviest weighted arrow as a primary interdependency in Figure 3-2(a). Access via roads and bridges is also essential to repair every other sector (e.g., drinking water, wastewater, dams, and reservoirs). For instance, following Hurricane Maria in 2017, critical energy and water supplies for the Island were unable to be transported from the port to areas in need for several weeks because of road failures (Government of Puerto Rico 2018). This is why the interdependencies between roads and bridges and drinking water and wastewater are illustrated in Figure 3-2(a) as secondary interdependencies with medium weight arrows. Also, as can be seen from this example, flood protection

Table 3-3. *Summary of Impacts to Transit and Aviation Infrastructure from Climate Change and Potential Remedial Actions.*

Infrastructure impact	Operations impact	Potential response
Increasing heat because of climate warming and urban heat island effects		
Increased rate of deterioration of materials	Service disruptions and delays	Change in materials selection; materials innovation to develop new, more heat-tolerant materials; mechanical cooling; increased maintenance
Rail line cracking or buckling because of rail expansion that cannot be constrained by the materials that support the track (e.g., rail ties and ballast)	Service disruptions and delays	Speed restrictions, service reductions, installing sensing, warning, advisory, and crew dispatching systems; planning for more frequent inspection/repair
Overheated electrical equipment such as substations, switchgears, and signals	Service disruptions and delays	Higher temperature rating or increased A/C for equipment
Overheated underground urban transit systems	Rider/worker health and safety, equipment failure, compounding because of overcrowding	Increased ventilation and A/C systems; personnel, equipment, and climate monitoring systems
Overheated transit or airport intermodal vehicles	Service disruptions and delays	New technology or adaptation of cooling systems
Failed air-conditioning systems in public transit/aviation buildings	Health and safety; service disruptions	Increased ventilation and A/C systems; personnel, equipment, and climate monitoring systems
Higher risk of incidents because of ineffective breaking or visual obstructions from increased vegetation/leaf fall	Health and safety; service disruptions	Vegetation management, plant low-maintenance vegetation as a buffer

(Continued)

Table 3-3. Summary of Impacts to Transit and Aviation Infrastructure from Climate Change and Potential Remedial Actions. (Continued)

Infrastructure impact	Operations impact	Potential response
Rail/road embankment failure; airport/transit infrastructure shallow foundation settlement because of thawing permafrost	Health and safety; service disruptions	Relocation, underground cooling/insulation systems, mechanical/structural stabilization
Decreased lift and thrust because of higher ambient temperature	Service disruptions and delays; requirement for increased runway lengths; increased turbulence and fuel consumption costs	Planes designed to fly, take off, and land more efficiently; altered flight paths and approach routes; improved navigation and control
Higher risk of incidents because of increased average wind speeds or extreme wind gust events	Health and safety; airline service disruptions; infrastructure damage	Air traffic control adaptation; increased plane stability
Increasing average extent of areas burned by wildfires impacting infrastructure	Health and safety; service disruptions; infrastructure damage	Increased monitoring and precautionary measures; emergency management planning; and fire barriers
Increasing temperature volatility and changing freeze–thaw cycles		
Reduce performance of materials vulnerable to the effects of thermal expansion	Health and safety, increased maintenance, service delays	Use of joints more resistant to expansion; change in materials selection; materials innovation to develop new, more heat-tolerant materials; increased maintenance

(Continued)

Table 3-3. Summary of Impacts to Transit and Aviation Infrastructure from Climate Change and Potential Remedial Actions. (Continued)

Infrastructure impact	Operations impact	Potential response
Rail line contraction resulting in breaking or separating (e.g., pull-aparts) because of cold temperature change	Service disruptions and delays	Speed restrictions, service reductions, installing sensing, warning, advisory, and crew dispatching systems; planning for more frequent inspection/repair
Increased ice on roads, tracks, airport runways, and supporting infrastructure	Worker health and safety; service disruptions and delays	Increased monitoring/inspection, heated runways, more frequent clearing
Increasing intensity and frequency of precipitation events		
Flooding of airport runways, taxiways, aprons, and other supporting infrastructure	Health and safety; the use of airport as a shelter or relief hub; service disruptions; access disruptions; infrastructure deterioration; hazardous material releases	Engineering controls for critical assets (e.g., raising or retrofitting); relocation; emergency contingency planning; improved drainage and runoff facilities; resilient materials; improved maintenance
Flooding of rail and associated infrastructure (e.g., signals and substations, including underground transit systems	Service disruption, malfunctions of track or signal sensors, loss of facilities; increased risk of hazardous material spills	Engineering controls for critical assets (e.g., raising or retrofitting); improved drainage; advisories, warnings, and updates to dispatch centers, maintenance crews; operational or route modification

(Continued)

Table 3-3. Summary of Impacts to Transit and Aviation Infrastructure from Climate Change and Potential Remedial Actions. (Continued)

Infrastructure impact	Operations impact	Potential response
Washouts and mudslides because of precipitation	Health and safety; service disruption, increased risk of hazardous material spills	Engineering controls to stabilize earth; operational or route modification
Increasing frequency of tidal flooding because of sea-level rise		
Flooding (overland and groundwater-based) and erosion of airport, roads, and supporting infrastructure	Health and safety for motorists; the use of airport as a shelter or relief hub; service disruptions; infrastructure deterioration; impeded ground access; undermined roadway bases and foundations	Engineering controls for critical assets (e.g., raising or retrofitting); community-scale flood protection; relocation; resilient materials; improved maintenance
Increasing intensity and frequency of extreme coastal flooding events		
Coastal road, railway, subway, and supporting infrastructure flooding (overland and groundwater-based)	Health and safety; destruction of infrastructure and infrastructure deterioration; service disruptions	Engineering controls for critical assets (e.g., raising or retrofitting); community-scale flood protection; relocation; resilient materials
Flooding of airport runways or infrastructure	Destruction of infrastructure and deterioration; service disruptions	Engineering controls for critical assets (e.g., retrofitting); community-scale flood protection; resilient materials
Intense wind and debris from storms can damage landing lights, navigation and radar installations, and communications equipment	Health and safety; destruction of infrastructure and infrastructure deterioration; service disruptions	Engineering controls (e.g., retrofitting), including shielding and change materials selection; increased maintenance

(a)

(b)

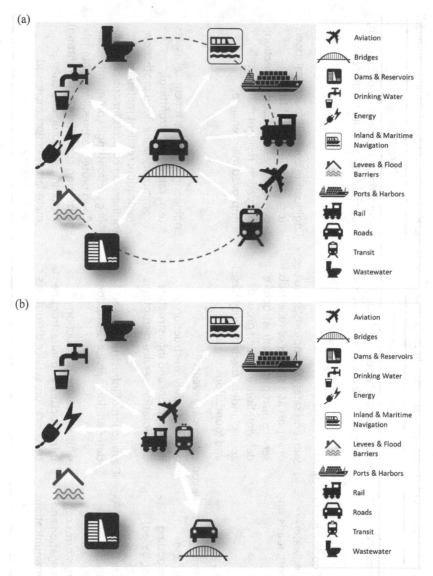

Figure 3-2. Sectoral interdependencies involving sectors: (a) roads and bridges, (b) transit and aviation. The arrow thickness indicates the level of criticality of sector interdependence. The double arrow indicates the dependence of both sectors on each other.

infrastructure is also essential and it, in turn, depends on an effective network of roads and bridges to provide for regular inspection and maintenance and to ensure efficient evacuation should there be a failure. This is a two-way dependency as the same infrastructure may protect critical sections of the highway from storm surge, for example.

Figure 3-2(a) also identifies various other sectors, such as transit, aviation, ports and harbors, navigation, and rail, that are dependent on the inflow and outflow of people and products via a well-maintained network of roads and bridges; any interruption could have significant social and economic impacts. Rail also provides critical benefits for roads and bridges as a major form of transportation of required raw materials for building and maintaining roads and bridges.

Figure 3-2(b) illustrates other transportation interdependencies related to transit and aviation. Transit and aviation infrastructure serve a critical role in the connection of goods, services, energy, and people and have the potential to impact other sectors while being directly and indirectly impacted. The relationship with the roads and bridges sector, with which transit and aviation shares much of its embodied infrastructure in addition to the ports and harbors sector, was previously described and is considered a primary interdependency for the transit and aviation sector. However, transit and aviation assets also form essential nodes and links within the labor, energy, and/or material networks that facilitate the functioning of all infrastructure sectors, including drinking water, wastewater, levees and other flood barriers, dams and reservoirs, ports and harbors, and navigation, as is illustrated in Figure 3-2(b). Therefore, because of its role in connecting the transit and aviation sectors to the other sectors, energy is considered a secondary interdependency. Other sectors are shown in Figure 3-2(b) as tertiary interdependencies that depend on the transit and aviation sector for efficient operation while also providing services essential for the continuous services that are required by the transit and aviation sector itself. This creates an opportunity for devising a holistic strategy that recognizes these complex relationships and potential emergent behaviors to prioritize resources as part of an overall resiliency strategy.

3.3 DRINKING WATER AND WASTEWATER

Drinking water and wastewater systems include infrastructure, operation of this infrastructure, environmental systems that provide water, and water itself. Drinking water and wastewater-related infrastructure include the following:

- Water storage facilities (both natural ones such as snowpack, landscape storage in watersheds, and groundwater systems and engineered storage in reservoirs, water tanks, aquifer recharge and recovery systems, and similar);
- Water distribution and wastewater collection networks (pipes, valves, canals, intake and outfall systems, etc.); and
- Water and wastewater treatment plants.

3.3.1 Impacts

Drinking water is essential for human life, and wastewater treatment is a key component of both public health and environmental protection. Maintaining the operation of both drinking water and wastewater systems is, therefore, critical to the operation of human settlements, in terms of both day-to-day functionality and extreme events where either system must be in a state of operational readiness or be rapidly re-established following failure.

Each component of the natural and engineered systems that provide us safe drinking water and allow safe disposal of wastewater is vulnerable to the impacts of our changing climate—including slow, creeping changes in average temperatures, water supply, and sea level, as well as to more episodic extreme events including floods, droughts, and wildfire. Water and wastewater systems are also vulnerable to the impacts of extremes in other sectors such as the following: transportation systems (Chapter 3.2) required to bring operators to and from plants and maintenance staff to and from component locations; energy (Chapter 3.1) that powers the treatment plants; and ecosystem impacts including sedimentation, saline intrusion, stormwater influx, and other forms of contamination, and system overload that can impact functionality.

Risks persists to the reliability and quality of water supplies as the climate changes (Chapter 2.1, 2.3, 2.5, and 2.6); to both natural and engineered storage as snowpack is reduced, spring runoff is accelerated, longer and deeper droughts increase the demand for stored water, and wildfire and drought increase reservoir sedimentation rates; and to the water treatment and distribution infrastructure as the climate becomes more variable, with more frequent extreme heatwave, drought, and precipitation events, all of which pose significant challenges to water quality, water treatment processes, and soil stability and the integrity of the pipe network.

Drinking water treatment plants are often located near water sources, such as rivers, increasing their vulnerability to extreme precipitation events. They are also vulnerable to events such as wildfires. Increased sediment and ash inflow can clog filtration systems and affect biological treatment processes. Large-scale urban fires, as was seen in Paradise, California, in 2018, can also result in toxic contamination of potable water pipe networks.

Wastewater treatment plants have similar vulnerabilities because of their proximity to rivers and coastal waters into which treated wastewater is discharged. Increases in extreme precipitation events, sea-level rise, and storm surge all pose acute challenges for these plants. In the older, eastern cities of the United States, sewers that collect both sewage and urban runoff (combined sewer overflow systems) pose additional risks to public health after extreme precipitation events because they increase the likelihood of release of untreated sewage into receiving water bodies. As for the piped water network, increased frequency and intensity of heatwaves, drought, and flooding will combine to increase stress on the wastewater piping network. Reduction in wastewater outflow during droughts may result in increased clogging of systems.

3.3.2 Status

As indicated in the 2017 ASCE report card on US infrastructure (ASCE 2017), the overall quality of drinking water in the United States is the safest in the world.

Most Americans receive drinking water from a community water system that is required to meet minimum national standards. In addition, conservation efforts are bearing fruit; nearly three-quarters of the major utilities in the United States have conservation programs in place and drinking water demand has remained flat since 1985 despite population increases.

However, both our drinking water infrastructure systems and the environmental systems that support water availability are under stress. Drinking water distribution systems in the United States are aging past their design lives. In particular, the distribution network is aging and is subject to frequent line breaks (ASCE 2017) but is being replaced at rates between a half and a quarter of what is needed to maintain its current functionality. It is estimated that leaking pipes are wasting 14% to 18% of treated water per day. At the same time, in several regions in the United States, surface and groundwater supplies are already stressed—by increasing demand, by declining runoff, and by declining groundwater recharge. Water quality is diminishing in many areas, particularly because of increasing sediment and contaminant concentrations after heavy downpours. These existing challenges will be further exacerbated by climate change, as described in Chapter 2, with more details provided in Table 3-4.

Wastewater systems are similarly currently strong but with critical weaknesses. Nearly 76% of the US population is now served by wastewater treatment plants. Years of plant upgrades, coupled with more stringent federal and state regulations, have improved water quality across the nation and significantly reduced untreated releases. Wastewater is increasingly being seen as a water supply opportunity, with increased investment in treatment and reuse. In addition, innovation in treatment methods and waste-to-power recovery hold promise in developing additional revenue streams for waste processing.

The challenges to the US wastewater systems lie in demand—it is estimated that 532 new systems will be needed by 2032 to meet treatment needs—and in environmental stresses and climate change. These are noted in Chapter 2 and Table 3-4. In particular, increased rainfall intensity will increasingly require decoupling of stormwater management from wastewater management. Sea-level rise will require a relocation of, or a rethinking on, coastal treatment plants.

3.3.3 Response

Although water and wastewater systems in the United States are exposed and sensitive to the likely impacts of a changing climate, building adaptive capacity into the system, such as through redundant water supplies for a community, can reduce the overall vulnerability of systems. A number of steps are available that we can take to further enhance the adaptive capacity and overall resilience of our national water and wastewater systems, and all these adaptions will provide benefit not just in the future but also in the present. A number of these potential adaptation actions are listed in Table 3-5.

Table 3-4. *Critical Climate Variables and the Risks They Pose to Water and Wastewater Infrastructure and Operations and Water Quantity and Quality.*

Critical climate variables	Risks to water and wastewater infrastructure and operations	Risk to water quantity and quality
Extreme rainfall events (heavy showers and long periods of rain)	Flooding of water and wastewater treatment plants in river valleys may threaten their ability to operate. Impacts on transportation networks may keep relief workers from reaching plants	Increased contamination, trash, and debris in urban runoff. Overloading of combined sewer overflow systems, with the potential release of sewage to public water sources, resulting in environmental and public health impacts. Contamination from flooding of areas with septic systems
Change in seasonal and annual average rainfall	Increased stormwater inflow to coupled stormwater-wastewater systems in wetter regions —	Decreased water availability in arid regions
Increase in seasonal and annual average temperature		Increased evapotranspiration, longer growing season resulting in increased plant water demand, increasing both agricultural and outdoor residential demand

(Continued)

Table 3-4. *Critical Climate Variables and the Risks They Pose to Water and Wastewater Infrastructure and Operations and Water Quantity and Quality. (Continued)*

Critical climate variables	Risks to water and wastewater infrastructure and operations	Risk to water quantity and quality
Sea-level rise and storm surge	Flooding of wastewater treatment plants along the coast, resulting in wastewater spills and/or damage to treatment plants Back-flooding of wastewater pipes, resulting in backflow into low-lying homes and neighborhoods (Grady 2014)	Saltwater intrusion into groundwater used for drinking Saltwater intrusion into surface waters at the potable water intake point
Maximum temperature and number of consecutive hot days (heat waves)	Desiccation of ground, which can increase stress on pipes and increase breaks, particularly when coupled with increased water demand	Increased water demand. Warmer water, reduced dissolved oxygen concentration, and increased potential for algal blooms
Drought (consecutive dry days) and aridification	Desiccation of ground that can increase stress on pipes and increase breaks, particularly when coupled with increased water demand. Particularly problematic in alternating drought/flood conditions	Decrease in water availability. Increase in water demand. Potential for dust- or ash-on-snow events that rapidly melt or sublimate snowpack, decreasing water storage and further decreasing supply. Increased reliance on groundwater, which lowers groundwater levels that, in turn, feeds back to further decrease surface-water supplies

(Continued)

Table 3-4. *Critical Climate Variables and the Risks They Pose to Water and Wastewater Infrastructure and Operations and Water Quantity and Quality.* (Continued)

Critical climate variables	Risks to water and wastewater infrastructure and operations	Risk to water quantity and quality
Reduced snowpack and/or earlier melt Aridification impacts on upland forests (increase in wildfire, infestations)	— Increased erosion, resulting in increased reservoir sedimentation, the potential to bury diversion infrastructure. Ash in runoff from wildfires threatens the ability of treatment plants to treat water. Extensive urban burns, resulting in toxic contamination of the pipe network	Decrease in water storage and supply Overall trend toward decreased supply in arid areas. Wildfire and infestations such as bark beetle decrease the ability of forests to maintain snowpack in spring, decreasing water storage and supply. Water quality impacts and increased sediment and ash from burn-area runoff

Table 3-5. Potential Adaptation Actions for Water and Wastewater Impacts.

Potential adaptation actions	How they will help?
Create protective infrastructure (e.g., berms and levees) to address the impacts of riverine flooding on water and wastewater treatment plants	Reduces risk of flooding of water and wastewater treatment plants because of extreme precipitation events, rapid snowmelt runoff, sea-level rise, or storm surge. However, this protective infrastructure may only buy time. If flood heights continue to rise, at some point the potential performance of protective structures may be exceeded
Disconnect urban drainage networks from the sewer system	Reduces the risk of raw sewage releases to receiving waters resulting from extreme precipitation events
Relocate or redesign/reengineer coastal wastewater treatment plants	Prevents flooding of treatment plants because of sea-level rise or storm surge
	Prevents sewage backflow into low-lying neighborhoods as the sewage collection network is flooded
Develop alternate potable water supplies with different failure modes (multiple surface-water and groundwater sources, desalinization, and water reuse)	Mitigates decreasing supplies because of increasing temperatures/aridification, longer and more intense droughts, and decreasing the ability to store water in upland forests
Perform thinning and prescribe burns in overgrown upland forests to prevent catastrophic wildfire. Reinitiate healthy fire into forest ecosystems	Reduces sedimentation and water quality impacts from wildfire. Mitigates decreasing water supply and access and decreasing water quality
Interconnect neighboring water treatment plants where feasible	Provides redundancy in the potable water system by allowing neighboring water treatment plants to share water in the event that one is temporarily taken off-line

(Continued)

*Table 3-5. Potential Adaptation Actions for Water and Wastewater Impacts.
(Continued)*

Potential adaptation actions	How they will help?
Where feasible, develop interconnected, modular wastewater treatment systems with different levels and types of exposure	Minimizes the potential for failure of both plants and maximizes the ability to maintain treatment capacity in the event that one plant is taken off-line
Increase funding at all levels—federal to local—for distribution network repairs and upgrades. Increase the rate of system upgrades nationally	Reduces system fragility and increases available water by reducing existing leakage in the potable water distribution network
Engineer more adaptive piping options to accommodate heat/drought/flood/demand stresses on system components	Has the potential to reduce the sensitivity of the existing distribution and collection systems to climate extremes
In all future infrastructure repairs or upgrades, consider the potential failure mechanism of the system component being repaired or replaced (both in today's climate and for likely future climate stressors) and build in redundancy using system components with other failure modes, to build resilience in the overall system	Builds in the engineering today to avoid preventable future failures
Create the funding streams needed to support building in redundancy to address future climate change and create the regulatory environment around that funding to assure it is applied as intended	Assures we do not miss opportunities to increase our adaptive capacity because of a lack of funding and awareness. Saves money in the long term
Assure that the legal system allows for innovation	Provides the flexibility needed to adapt operations and infrastructure to changing conditions and demands

3.3.4 Interdependencies

With their frequent location near watercourses or coastlines, water and sanitation are exposed networks with potential for high societal consequences if disrupted (Bierkandt et al. 2015, Chang 2016) and that are understudied with respect to repetitive exposure. Even brief unplanned interruptions (>1 day, <weeks) to

water supply, in addition to energy supply, can compromise emergency services and businesses (Brozović et al. 2007). The cumulative cost to the local economy could be high if persistent disruptions prompt a business to relocate and, thus, remove employment opportunities. In short, the connectedness of everyday life exacerbates our vulnerability and exposure to less extreme events, through the secondary and even tertiary impacts of flood disruption.

A key interdependency involves the dependence of the drinking water and wastewater sectors on the energy sector (highlighted as a primary interdependency in Figure 3-3). As previously mentioned, the drinking water, wastewater, and energy sectors are critical because the short-term impacts of any disruption within these sectors can have major consequences. Therefore, uninterrupted operation and functionality of drinking water and wastewater infrastructure is critical and depends foremost on energy supply, particularly with regard to the operation of dams and reservoirs, which is also illustrated as a primary interdependency in Figure 3-3. Any disruption in the efficient operation of either the energy or dam and reservoir sectors can have an immediate and disastrous impact on access to sufficient water to meet the various water supply demands.

Another critical interdependency (illustrated as a secondary interdependency in Figure 3-3) within the drinking water and wastewater sectors involves continuous access via roads and bridges in the event of any repairs to maintain continuous operation. As previously mentioned, critical water supplies for the

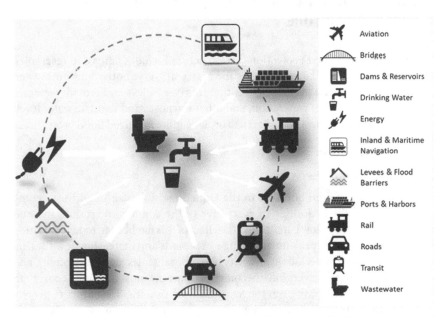

Figure 3-3. Sectoral interdependencies involving the drinking water and wastewater sectors. The arrow thickness indicates the level of criticality of sector interdependence. The double arrow indicates the dependence of both sectors on each other.

island of Puerto Rico following Hurricane Maria in 2017 were unable to be transported from the port to areas in need for several weeks because of road failures (Government of Puerto Rico 2018).

Facilitation of efficient drinking water and wastewater services is also dependent on the functionality of transit and aviation, which form essential nodes and links within the labor, energy, and/or material networks. In turn, efficient operation of transit and aviation, as well as the navigation and ports and harbors sectors, requires uninterrupted access to drinking water and wastewater. Therefore, all transit and aviation sectors, including rail, are illustrated in Figure 3-3 as codependencies, whereas the dependence of the navigation and ports and harbors sectors is also illustrated as a tertiary interdependency.

Drinking water and wastewater also depend on flood protection infrastructure for effective and continuous operation. Potential impacts can include loss of power, damage to infrastructure and other assets, and danger to personnel. For example, when the Geneva Water Works in Geneva County, Alabama, experienced power outages because of a major flood (EPA 2014), substantial sewer backups occurred, which released raw sewage and other toxic materials into the floodwaters. Such an event presents a major hazard to those who come into contact with floodwaters as well as those who depend on drinking wells for their household water supply.

3.4 WATER STORAGE AND FLOOD PROTECTION INFRASTRUCTURE

Water storage and flood protection infrastructure includes dams and reservoirs, levees, and other flood barriers. Dams and reservoirs serve both water storage and flood protection functions for large sections of river valleys and provide water for recreation, hydropower generation, and other purposes and regulate water levels for navigation; levees and flood barriers provide more localized flood protections.

3.4.1 Introduction

3.4.1.1 Dams and Reservoirs

Three centuries of dam building in the United States has resulted in sweeping changes to riverine hydrology in every area of the country to retain water and regulate downstream flow conditions. The effect of this has been to reduce variation in flows by decreasing peak flows during wet periods and increasing flows during otherwise lower-flow periods at a variety of temporal scales (daily, seasonally, and annually). As a result, rivers have become "perhaps the most impacted ecosystem on the planet as they have been the focus for human settlement, and are heavily exploited for water supply, irrigation, electricity generation and waste disposal" (Malmqvist and Rundle 2002, p. 134).

Following the arrival of Europeans in the United States, dam construction began as a means of powering mills for grinding grain, and subsequently regulating rivers for navigation. The US Army Corps of Engineers (USACE) was first authorized to improve navigation on the Ohio and Mississippi Rivers with the General Survey Act of 1824 and used a combination of wing dams and dredging to alter flow conditions. The first cement lock and dam structure constructed on the Ohio River was the Davis Locke and Dam near Pittsburgh, Pennsylvania, which opened in 1885 (USACE 2013).

Large-scale construction of dams for water supply, hydropower, and flood control, particularly in the arid West, began with the vision of John Wesley Powell and led to the establishment of the Reclamation Service (now Bureau of Reclamation) in 1907 (Reisner 1993) and the construction of the first major multipurpose dams: Roosevelt Dam (Salt River) and Wilson Dam (Tennessee River) (Billington et al. 2005). The US government funded a flurry of dam construction in the mid-twentieth century with the goals of providing water and power for a quickly growing population west of the Mississippi River. These dams and reservoirs have increased water supply reliability in this arid region, reduced flood risk, provided hydroelectric power, served as recreation destinations, and created economic opportunity.

Dams and their operation for downstream flow regulation have caused significant ecological changes to downstream riparian and aquatic ecosystems, including fragmented and degraded river ecosystems, declining fish stocks, and impounded contaminants; how to balance the needs of aquatic and riparian ecosystems and humans remains one of the most important questions of our time (Nilsson and Berggren 2000). As a result, dam removal has become an increasingly common method of river restoration in the United States, with more than 1,185 documented projects taking place in the last century (American Rivers 2014). These projects have varied in size and complexity from low-head, run-of-river structures in coastal streams in Massachusetts to massive dams along the Elwha River in Washington.

3.4.1.2 Levees and Flood Barriers

Levees are linear earthen barriers set parallel to a stream that are designed to prevent floodwaters from leaving the stream channel and entering all or part of the adjacent floodplain, or otherwise contain, control, or divert the flow of water to reduce risk from flooding. Levees may be engineered to prevent flooding from floods of a known magnitude or they may be no more than *push piles* of sediment (spoil banks) with unknown engineering and flood-retention properties. They also function as part of water storage and management systems in some parts of the country. Other types of flood barriers include berms, floodwalls, and temporary barriers.

NFIP defines a levee as "a man-made structure, usually an earthen embankment, designed and constructed in accordance with sound engineering practices to contain, control, or divert the flow of water to reduce risk from temporary flooding" [44 CFR (Code of Federal Regulations) 59.1]. A levee may be a standalone structure or may be part of a system of structures for flood risk reduction in an area. The NFIP regulations define a levee system as "a flood protection system that consists of a levee, or levees, and associated structures, such as closure and drainage devices, which are constructed and operated in accordance with sound engineering practices." A spoil bank is not considered a levee under NFIP (44 CFR 65.10).

Levees, when performing as intended, reduce the risk of flooding for floodplain communities by detaining water in the stream channel that would otherwise have overflowed the floodplain. A levee breach or failure can be catastrophic, with the resultant flooding causing loss of life, damage to property, and economic disruption. Poorly constructed levee systems may trap floodwaters behind the levees or may so increase floodwater height in the channel so as to increase flood risk to nearby communities where levees may be lower or absent.

3.4.2 Impacts

3.4.2.1 Dams and Reservoirs

Changes anticipated as a result of climate change will affect water supply and hydropower production, flood protection, and recreation, as described in the following:

- Water supply (for municipal industrial and agricultural uses) and hydropower production: Changes in the flow regime could affect reliability through reductions in available water supply (seasonal or multiyear drought; increased evaporation; a shift in freshet timing, resulting in lower base flows during dry seasons; and loss of headwaters glaciers) and through changes in operations to manage flood risk or water supply. Demand for water and power may increase in the summer months because of increased cooling demand. In watersheds where precipitation is projected to increase (more precipitation and increased glacial melt), water supply and hydropower conditions may become more favorable.

- Flood protection: Changes in precipitation form, intensity, magnitude, and seasonality are projected for many watersheds; increases in the magnitude of extremely high runoff events are likely, at least seasonally, in many parts of the United States. Coupled with changes in land use and urbanization, these changes are likely to significantly alter flood risk for floodplain communities (Wehner et al. 2017). These projected changes are likely to affect flood operations at dams, including the magnitude and frequency of flood releases, and the operation of other flood infrastructure (such as floodways). Reservoir capacity for flood risk management may be reduced if sedimentation rates increase in response to increased runoff and in response to increased water storage demands for hydropower and irrigation/water supply purposes.

- Recreation: Higher temperatures could see higher recreation dependency; in areas where water supply decreases, water quality may become too impaired in some seasons for some or all recreational activities and reductions in water quantity may impair recreational use; in areas where water supply increases, recreational infrastructure may be flooded and recreational opportunities curtailed.

In addition, there are tradeoffs and uncertainties related to the use of dams and reservoirs because they are used both for water supply storage and for flood protection.

- For a given reservoir, water storage volumes are finite and, through sediment influx, decline over time. Capacity may be increased by raising the dam or maintained by passing or dredging sediment, and these may be cost-effective in some circumstances. Reservoirs are typically divided into *pools* that define the amount of space set aside for each authorized purpose. Because capacity is finite, changing the volume of one pool necessitates changes to one or more of the other pools. Pool volumes may vary on a seasonal schedule that reflects the current seasonal balance between the space required for flood storage and all other uses.

- Climate change is projected to increase the magnitude and frequency of precipitation extremes (floods and droughts), while reducing the frequency of average precipitation days. Thus, for many parts of the United States, both flood and drought risks are projected to increase. Mitigating water supply by increasing water storage capacity would require reductions in flood storage capacity or in pools for other authorized purposes; likewise, increases in storage capacity for floods will decrease the water available for other uses.

- Because current-generation climate models have a lot of uncertainty in their precipitation projections, which is compounded by the process of translating these estimates into stream flows using hydrologic models (Clark et al. 2016), it is not feasible to model the water supply–flood risk tradeoff with the accuracy needed to estimate flood storage capacity volumes for the purpose of reallocating reservoir storage. In addition, increases in precipitation extremes (amount, intensity, and spatial distribution) are in general linked to higher runoff and soil erosion rates, and higher stream sediment transport rates (Li and Fang 2016). Future land use and land cover will also be important factors, and both are also poorly constrained in model projections. Thus, future capacity loss rates because of reservoir sedimentation are an important unknown.

- Currently, in portions of the United States, precipitation variability plays a significant role in flood risk: Historically, many droughts have abruptly ended with extreme precipitation events (Dettinger 2013, Maxwell et al. 2017). Under the greater projected future extremes, storing additional water in the flood pool to mitigate drought may result in difficulties evacuating that water when the drought ends, thereby increasing downstream flood risk.

Impacts that may result in water storage and flood protection systems resulting from changes to critical climate variables are listed in Table 3-6.

Table 3-6. *Critical Climate Variables and the Risks They Pose to Water Storage and Flood Protection Dams and Reservoirs, and Their Operations.*

Critical climate variables	Risks to water storage infrastructure and operations, recreation, and hydropower generation	Risk to flood protection infrastructure
Extreme rainfall events (heavy showers and long periods of rain)	More water storage and hydropower generation. Requires more careful management to make sure appropriate amounts are stored and released. Recreation and water quality may be impaired by increased pollutant loads. Because of flooding, recreational facilities at reservoirs may be closed or damaged	Increased flood volumes; longer flood durations; flashier flood flows; more frequent flood events; increased sedimentation impacting flood pool capacity, particularly if extreme rainfall events follow drought so initial reservoir levels are low. Risk of exceeding flood storage capacity; risk of overtopping dams and spillway failures. Risk of releases exceeding downstream channel capacity. Risk of dam failure, cascading infrastructure damage, and loss of life
More rapid snowmelt runoff, and the likelihood of rain-on-snow events	Requires more careful management to make sure appropriate amounts are stored and released	Risk of exceeding flood storage capacity; risk of overtopping dams and spillway failures. Risk of releases exceeding downstream channel capacity

(Continued)

Table 3-6. Critical Climate Variables and the Risks They Pose to Water Storage and Flood Protection Dams and Reservoirs, and Their Operations. (Continued)

Critical climate variables	Risks to water storage infrastructure and operations, recreation, and hydropower generation	Risk to flood protection infrastructure
Change in seasonal and annual average rainfall	Decreased water storage in water supply reservoirs in arid regions, with a commensurate decrease in hydropower generation and recreation potential	Season of greatest flood risk may change; advance in the timing of spring runoff flooding
Increase in seasonal and annual average temperature	Increased evapotranspiration, longer growing season resulting in increased plant water demand, increasing demand for water stored in reservoirs	Increased winter temperatures may result in higher winter stream flows and a commensurate need for flood risk management
Sea-level rise and storm surge	May cause saltwater intrusion into stored water supply	Rise in the base level may limit downstream channel capacity and, therefore, limit flood release volumes
Maximum temperature and number of consecutive hot days (heat waves)	Increased water demand. Warmer water, reduced dissolved oxygen concentration, increased potential for algal blooms. Increased demand for reservoir recreational facilities	No effect

(Continued)

Table 3-6. *Critical Climate Variables and the Risks They Pose to Water Storage and Flood Protection Dams and Reservoirs, and Their Operations. (Continued)*

Critical climate variables	Risks to water storage infrastructure and operations, recreation, and hydropower generation	Risk to flood protection infrastructure
Drought (consecutive dry days) and aridification	Decrease in water availability and hydropower generation. Increase in water demand. Potential for dust- or ash-on-snow events that rapidly melt or sublimate snowpack, decreasing water storage and further decreasing supply. Increased reliance on groundwater which lowers groundwater levels, which, in turn, feeds back to further decrease surface-water supplies. Decreases in water levels in the reservoir may leave boat ramps above water levels and other recreational facilities far away from the water	Drought reduces flood risk, but very commonly in at least portions of the United States, droughts end with extreme precipitation events: from 1950 to 2013, 33%–40% of California droughts, and 60%–74% of all Pacific Northwest droughts ended by landfalling atmospheric river events (Dettinger 2013); 73% of southeastern US droughts during the 1979–2013 warm season also ended with large precipitation events (Maxwell et al. 2017)
Reduced snowpack and/or earlier melt	Decrease in water storage and supply, hydropower generation, and recreational potential	May require a change to reservoir operations to account for a change in the timing of flows

(Continued)

Table 3-6. Critical Climate Variables and the Risks They Pose to Water Storage and Flood Protection Dams and Reservoirs, and Their Operations. (Continued)

Critical climate variables	Risks to water storage infrastructure and operations, recreation, and hydropower generation	Risk to flood protection infrastructure
Increased flashiness	Increased contamination in the runoff, especially in urban areas. Increased debris inflow to reservoirs	Changes to the location of sediment accumulation in reservoirs as reservoir levels rise and fall, with potentially negative implications for flood pool capacity
Aridification impacts on upland forests (increase in wildfire, infestations)	Increased erosion, resulting in increased reservoir sedimentation, potential to bury diversion infrastructure. Overall trend toward decreased supply, and decreased hydropower reliability, in arid areas. Wildfire and infestations such as bark beetle decrease the ability of forests to maintain snowpack in spring, decreasing water storage and supply. Water quality impacts and increased sediment and ash from burn-area runoff. Decrease in water available for reservoir and riverine recreation	Postwildfire flood flows produce more runoff for a given amount of precipitation, and these runoff flows are bulked with debris and sediment, resulting in higher reservoir inflow volumes and risk to infrastructure from debris

3.4.2.2 Levees and Flood Barriers

Exposure to the forecasted atmospheric conditions, based on current climate change trends, are unfavorable and may cause levees, primarily earthen levees, to weaken through strength reduction, drying, soil desiccation cracking, shrinkage, microbial oxidation of soil organic matter, fluctuation in the groundwater table, significant land and surface erosion, and highly dynamic pore-pressure changes (ASCE CACC 2015, Bates and Lund 2013, Brooks et al. 2012, Dunbar et al. 2007, Dyer et al. 2009, Port and Hoover 2011, Vahedifard et al. 2020, Vicuña et al. 2006). These activities affect the soil's strength properties and the structural mechanics within the body of the levee. The effect on soil strength properties is primarily related to changes in soil moisture and suction within the unsaturated levee body. Table 3-7 summarizes major climate change features and their effects on earthen levees. A more detailed discussion is provided in the following paragraphs.

Possible changes in precipitation include variation in total rainfall amounts as well as an increased occurrence of extreme events, such as intense rainfall, flood, or drought. An increase in the total precipitation will increase the level of saturation within the unsaturated zone and may decrease the depth to the water table. Resulting increases in pore pressure may decrease suction, lowering the shear strength of soil and possibly resulting in failure (Dehn et al. 2000, Lee and Jones 2004, Clarke et al. 2006). Conversely, a decrease in total precipitation will lower the level of saturation, increasing the soil's effective strength through higher suction. However, extended drought conditions may result in a loss of these improvements because of excessive soil drying, resulting in a decreased contribution from suction, desiccation cracks, heavy shrinking, and loss of organic matter (Robinson and Vahedifard 2016).

High precipitation intensity can have a significant negative effect on natural slopes and geotechnical structures. Sudden increases in saturation will reduce the effect of suction, thus lowering the effective strength of the soil (Lu and Likos 2004). In addition, intense rainfall often causes erosion of surface materials and has also been associated with soil piping within slopes (Jones 2010). These processes may cause or enhance the failure of natural and built slopes (Hungr et al. 2005, Iverson et al. 2011). Furthermore, the effects of intense rainfall can be enhanced if preceded by an extended period of drought (Vahedifard et al. 2016). In addition, models have illustrated that under partially saturated conditions, even nonextreme above-average rainfall can result in slope failures (Leshchinsky et al. 2015). Because of this, expected changes in precipitation occurrence are significant, as most geotechnical infrastructure, including levees, in the United States is designed using intensity–duration–frequency (IDF) curves established under the stationarity assumption (i.e., statistics of extreme events will not vary significantly over a long period of time). Because climate change is anticipated to influence precipitation patterns, the resulting changes in the statistics of extreme events may render the use of stationary IDF curves and previous designs ineffective for the mitigation of risk (Vardon 2015).

Table 3-7. Potential Impacts of Climate Change on Earthen Levees.

Climate change feature	Fundamental impact	Practical impact
Increased temperature	Higher evaporation rate/soil drying, SOC oxidation, changes in vegetation amount, snow, ice, and permafrost melting	Desiccation cracking, shrinkage, land subsidence, reduced strength of artic soils, the release of entrapped carbon, increased risk of mass wasting at higher elevations
Decreased mean precipitation	Soil drying and water table lowering, vegetation reduction	Possible desiccation cracking and shrinkage, loss of cover, and increased risk of erosion
Increased mean precipitation	Soil wetting and water table rise	Decreasing suction leading to reduced shear strength
Drought	Extreme soil drying and water table lowering	Significant desiccation cracking and shrinkage and increased susceptibility to intense erosion because of increased permeability from cracking and shrinkage
Intense precipitation	Rapid soil wetting, overland flow	Sudden changes in suction possible leading to heightened failure risk, possible erosion, and mass wasting
Flood/sea-level rise	Large pore pressure increases, soil wetting	Lowered suction within flood protection infrastructure because of wetting, increased risk of multiple failure mechanisms such as piping, overtopping, and erosion

Rainfall-triggered instabilities in levees are analyzed primarily using extreme precipitation estimates, derived using the so-called stationarity assumption (i.e., the statistics of extreme events will not vary significantly over a long period of time). However, extreme precipitation patterns have been shown to vary substantially because of climate change, leading to unprecedented changes in the statistics of extremes, a notion known as nonstationarity. It is shown that the usage of stationary rainfall data can lead to underestimations in the hydromechanical behavior of natural and human-made earthen structures (Robinson et al. 2017, Vahedifard et al. 2017b, Ragno et al. 2018). The findings highlight the importance of site-specific assessments to quantify the potential impacts of climate change on the performance of current and future earthen structures.

Extended precipitation events, both extreme and moderate, can cause flood conditions. Floods pose similar risks to extreme precipitation with increased levels of erosion, soil wetting, and high pore pressures in levees (Jasim et al. 2020, Vahedifard et al. 2020). Sufficiently high water levels may cause overtopping of levees, causing a significant external erosion of the earthen structure. The possible changes in precipitation event occurrence may change the rate of flood occurrence, rendering existing flood protection infrastructure and planning insufficient (Vardon 2015). In addition, the increased pore pressures may cause piping and other forms of failure to occur (Jasim et al. 2020, Vahedifard et al. 2020). Similar risks are posed by sea-level rise to levees and other coastal infrastructure. However, the effects of sea-level rise will occur gradually but will not recede like floodwaters steadily, increasing the risk along coastal communities.

Although the expected changes in precipitation and weather patterns are projected based on changes in global temperatures, these increasing temperatures exercise their own influence on earthen levees. Higher temperatures will accelerate evaporation and soil organic carbon (SOC) oxidation (Davidson and Janssens 2006, Conant et al. 2011). Increased evaporation rates, combined with precipitation changes, will increase the frequency of pore-pressure cycling, which has been known to cause strain softening and changes in permeability (Kovacevic et al. 2001, Potts et al. 1997, Nyambayo et al. 2004). In addition, increased evaporation rates because of high temperatures may exacerbate the development of negative effects because of drought conditions. Oxidation of SOC may cause land subsidence in highly organic peat soils and may cause an increased rate of sea-level rise in some regions. SOC oxidation accounts for approximately 75% of the elevation loss because of peat subsidence, whereas the remaining 25% is attributed to secondary consolidation and compaction of organic soils (Mount and Twiss 2005). Further, this oxidation is not accounted for in current emission estimates (Scharlemann et al. 2014). In addition to these effects, temperature increases cause the melting of permafrost in arctic regions, greatly reducing soil strength as well as melting ice and snow at high elevations, thus increasing the risk of mass wasting (NASEM 2016, Gariano and Guzzetti 2016).

3.4.3 Status

3.4.3.1 Dams and Reservoirs

The National Inventory of Dams (NID) lists more than 91,400 federal, state, public, and private dams with high or significant hazard potential or those that meet minimum size criteria (USACE 2019a). These dams are scattered around the country on many of the major waterways of America. In addition, there are an estimated 2.6 million small impoundments in the conterminous United States, most with a surface area of less than 10,000 m² (107,640 ft²) (Smith et al. 2002, Renwick et al. 2005). Most of these dams are low-head dams, defined as not reaching a height of 8 m (25 ft) (Pohl 2002). The condition of many of these dams is unknown, and some may pose significant hazards to downstream communities and infrastructure if they breach.

Although NID lists some 4,000 dams constructed in the United States since 2000, the majority of these have been smaller dams with small reservoirs. The construction of new, high dams with large reservoirs has been virtually halted because of a lack of public support, a lack of evidence of destructive effects on local and downstream ecology, and an overall lack of undammed sites that can produce adequate resources such as water storage and electricity generation (McCully 2001, Pohl 2002). Public perception of dam construction has also been affected by an acrimonious history of acquiring land by eminent domain to build large reservoirs (Knight and Rummel 2014).

The average age of dams in NID is 57 years, and 5,923 dams for which completion dates are known are more than 100 years old. Many of the existing dams in the United States have outlived their original purposes and have remained in service past their design structural lifetime (Gulliver and Arndt 1991). Consequently, dam condition and dam failure are an increasing concern where the magnitude of future floods may increase: As dams age, their structural integrity can become compromised by issues such as piping (water forcing its way through the bottom of earthen dams) and cracking (water pressure creating small openings in concrete dams). In addition, most dams were designed for historical flow conditions and might not be correctly sized relative to future flows.

Approximately 34% of all US dam failures have resulted from overtopping, 30% from foundation defects, 20% by piping, and the remainder by cracking and inadequate maintenance and upkeep (ASDSO 2019b). These factors resulted in 173 dam failures and 587 incidents where dam failure would have occurred without intervention just in the period from January 2005 through June 2013 (ASDSO 2019b). High-profile recent examples include the loss of Spencer Dam, Nebraska (2019), and the near-failure of Oroville Dam, California, in 2017 because of erosion of the emergency spillway. Currently, more than 2,170 deficient, high-hazard potential (potential for loss of life or significant property damage) dams are present in the United States (ASDSO 2019a).

Dam ownership in the United States is highly variable: 63% of NID-listed dams are privately owned, 7% are state-owned, 4% are owned by the federal government, and the remainder are owned by local governments and public utilities (USACE 2019a). Three-quarters of dams are regulated by state (69%) or federal (5%) agencies, leaving one-quarter of NID structures that are regulated by neither agency type. Regulation of the remaining inventory of approximately 2.6 million small dams is unclear.

Despite the plurality of ownership, dams may be operated as a system to achieve water resource goals. For example, Section 7 of the Flood Control Act of 1944 and other laws allow USACE to operate non-USACE reservoirs, locks, dams, and other water control projects for the purposes of flood risk management. The procedures for the operation of dams and associated infrastructure (spillways, floodways, and levees) as a flood risk management system for a given watershed are laid out in water control manuals published by USACE. These manuals are regularly updated, and a well-established procedure exists for any deviations from these procedures to enable a flexible response to situations as they arise. Flood operations are coordinated among federal, state, and local governments and utilities. USACE, FEMA, and other federal agencies also coordinate flood risk management activities with state, tribal, and local agencies through programs like the state Silver Jackets. Similar coordination exists among entities that manage rivers for water supply, navigation, and other purposes (e.g., water levels for navigation on the Mississippi and Ohio Rivers are managed by USACE).

Dam safety is an important federal concern. FEMA serves as the lead federal agency for establishing and reviewing dam safety guidelines for all federal dams and dams regulated by federal agencies (FEMA 2004a). FEMA is the lead agency for the National Dam Safety Program, which works with federal and private sector partners on dam safety technology and provides financial assistance to state dam safety programs and also training and public awareness (FEMA 2015). High-hazard-potential dams, which are dams whose failure or misoperation could result in loss of human life and significant economic, environmental, or other losses (FEMA 2004b), are encouraged to have emergency action plans that coordinate disaster response among federal, state, and local governments (FEMA 2013). Each federal dam-owning agency has a dam safety program that ensures that its structures are properly maintained and inspected. The USACE Inspection of Completed Works program ensures that federally funded dams, nonfederally owned dams, and other flood risk management infrastructure are properly maintained to provide expected flood damage reduction benefits and retain eligibility for federal funds to repair structures if they are damaged in a natural disaster. FEMA dam safety guidelines do not currently require the consideration of projected climate change impacts to structures or operations.

Although federal dam safety guidelines are increasingly coordinated, dam safety regulations are more variable at the state and local levels. States are responsible for regulating dams within their boundaries, and state programs vary in the types of dams they regulate, their ability to enforce existing law, and available funding for these efforts (ASDSO 2019a). Most dams in the NID are privately owned, with the owner solely responsible for dam safety, maintenance, and repair (ASDSO 2019a). Of the privately owned dams in the 2018 NID, 6,167 are categorized as high-hazard potential (USACE 2019a).

3.4.3.2 Levees and Flood Barriers

More than 30,000 documented miles and up to an estimated 100,000 mi of levees form a key component of the nation's critical infrastructure (NRC 2012, ASCE 2017). These levees protect millions of people, valuable property, and critical infrastructure across the country (NRC 2012). They also function as a part of water storage and management systems in some parts of the country (Vahedifard et al. 2016). The documented levees in the USACE Levee Safety Program protect more than 300 colleges and universities, 30 professional sports venues, 100 breweries, and an estimated $1.3 trillion in the property (ASCE 2017).

Most of the levees across the country were built in the middle of the last century by federal, state, and local agencies or by private property owners. The average age of levees in the United States is 50 years and many are showing their age (ASCE 2017). Although there are newer or reconstructed levees, a large number of levees were built in response to the widespread flooding on the Mississippi River in 1927 and 1937 and in California after catastrophic flooding in 1907 and 1909.

Levee failure mechanisms are similar to those for dams and include overtopping, seepage and piping, erosion of the waterside face, and damage from flood debris (USACE 2019b). Levee safety is a topic of increasing concern at the federal level, resulting in the establishment of the National Committee on Levee Safety in 2007 and the development of the USACE-led National Levee Safety Program (USACE 2019b). Although levee construction has focused on containing flood risk from riverine flooding, urban flooding behind levees and other flood risk management infrastructure is also becoming a major problem in some portions of the country as a result of both increases in precipitation during extreme weather events and increasing urbanization.

There is an important synergy with sea-level rise for communities currently in or anticipated in the future to be within the tidal reaches of streams. Sea-level rise is likely to increase stream elevations in tidal reaches, raising floodwater elevations and impeding flows, all of which may contribute to reductions in levee performance in tidal reaches. In some areas, upstream dams have trapped sediment, resulting in subsidence in coastal regions, contributing to levee performance reduction. These factors may come together to increase flooding when a large storm hits. A recent example of this kind of compound event (Wahl et al. 2015) was the flooding in Houston, Texas, in 2017, where a combination of extreme precipitation, urbanization/inland flooding, and storm surge all played significant roles (Wahl et al. 2018, Zhang et al. 2018). Urbanization not only increased the runoff that overwhelmed flood defenses but may have also contributed to the extreme storm rainfall totals over the city (Zhang et al. 2018).

Records of previous events show that levee failures can lead to catastrophic damage to the economy, infrastructure, and population in the affected areas (NRC 2012). Despite the critical role that the levees play, the majority of the nation's levees are more than 50 years old and are rated to be in *poor condition*; their current status can be described as "mostly below standard, with many elements approaching the end of their service life. Condition and capacity are of serious concern with strong risk of failure" (ASCE 2017). The integrity of levees warrants even further attention considering recent climatic trends such as rising sea levels, increased rainfall intensity, drought, and other extreme weather events (Bennett et al. 2014, Jasim et al. 2017, Robinson and Vahedifard 2016, Vahedifard et al. 2015, 2017a, b). Catastrophic levee failures after the 2005 Hurricane Katrina demonstrated how the increased frequency of extreme events because of climate change has increased the risk toward levees (Dupray et al. 2010). As development continues to encroach in floodplains along rivers and coastal areas, an estimated $80 billion is needed in the next 10 years to maintain and improve the nation's system of levees (ASCE 2017).

According to the USACE's National Levee Database (NLD), levees are found in approximately 35% of the nation's counties, with nearly two-thirds of Americans living in a county with at least one levee. Earthen embankments make up 97% of all the levees in the USACE Levee Safety Program, whereas floodwalls make up the remaining 3%. The NLD contains 11,900 individual levee systems accounting for the nearly 30,000 mi of documented levees. The USACE maintains authority of more than 13,700 mi, whereas other federal, state, or local agencies are responsible for the remaining 15,400 mi in the NLD. Because of the large inventory of levees outside of the USACE's authority, the condition of the nation's levees is largely unknown, but future efforts are planned to gain a better understanding of the nation's levees, as authorized in the Water Resources Reform and Development Act of 2014. The USACE has performed engineering inspections and risk assessments to understand the condition and characterize the flood risk associated with levees in their authority. Currently, the USACE has completed risk assessment on more than 1,200 levee systems out of the 2,500 in the USACE program. The risk assessment shows that of USACE-owned levees, 5% are high to very high risk, 15% moderate risk, and 80% low risk. The assessments are based on several criteria, including possible loading events such as floods, storms, and earthquakes; the level of performance; and consequences of failure. Major deficiencies include culverts, seepage—the biggest risk driver—and vegetation. The numbers of high- and moderate-risk levees are expected to grow as more inspections are performed, raising awareness of their conditions. Currently, less than half of the levees in the USACE's authority have risk assessment and risk characterization studies performed on them.

3.4.4 Response

Dam and reservoir operations are fairly responsive to precipitation variability under current river management practice (water control manuals, deviations, drought contingency plans, and operating rivers as multiuse systems). Regular reassessment of changes to hydrometeorological conditions and extremes and improved seasonal river flow forecasts are necessary to maintain flexibility and improve water management. National policies to set dam safety and maintenance standards under changing climate are needed.

Flood management is moving away from relying mainly on hard infrastructure, such as levees and dams, and increasingly relying on a portfolio of nonstructural measures, such as building elevation and floodproofing, structure relocation, land acquisition, flood early warning systems, and land-use regulation.

A number of steps are available that we can take to enhance the resilience of our national water storage and flood protection systems, now and in the future. A number of these potential adaptation actions are listed in Table 3-8.

3.4.5 Interdependencies

The primary sector on which dams and reservoirs are dependent is energy. As with other sectors, maintaining an uninterrupted supply of power is key to the continuous operation of dams and reservoirs, particularly in those areas where hydropower is not an option for fulfilling energy needs. In those areas where hydropower is an option and is the major source of energy, a codependence exists where a continuous supply of energy is dependent on the proper functioning of the dam and reservoir system.

The continuous operation and functionality of drinking water and wastewater infrastructure is another critical need that depends not only on energy supply but also on the efficient operation of dams and reservoirs and is, therefore, also illustrated as a primary interdependency in Figure 3-4(a). Any disruption in the expected operation of a dam and reservoir system can have an immediate and disastrous impact on access to sufficient water to meet the various water supply demands.

The integrity of levees and other flood barriers directly affects several infrastructure systems that they are designed to protect. These include transportation infrastructure, power systems, and water and wastewater treatment systems. With regard to the energy sector, major flood events can result in widespread power outages that will have major impacts on all other sectors; the existence of flood protection infrastructure near power stations located in flood-prone areas is key to preventing such events. As discussed in the previous section, drinking water and wastewater infrastructure also depend on flood protection infrastructure for full and efficient operation, which includes the prevention of power loss, prevention of infrastructure damage, and danger to personnel because of an extreme event. Any flood event that causes power outages may result in substantial sewer backups and dangerous conditions for those who come into contact with floodwaters as well as those who rely on wells for their drinking water needs.

An example of a major event during which sufficient flood protection infrastructure would have had a drastic impact occurred in 2011 when Japan was struck by the Great Tsunami. The tsunami rose up to 10 m (33ft) above the sea level along a 500 km (311 mi) Pacific Coast and reached over the protective levees (Mori et al. 2012). The disaster caused more than 18,000 fatalities and a world record-breaking economic loss (Shimozono and Sato 2016). While coming up with procedures to mitigate the effects of flooding such as the construction of levees and flood barriers, the integration between different community members should also be considered. Previous studies (Bertin et al. 2014, Lee et al. 2017)

Table 3-8. *Potential Adaptation Actions for Dams and Reservoirs, Levees, and Flood Barriers.*

Potential adaptation actions	How they will help?
Complete levee mapping as outlined in the National Flood Insurance Program reform bill, and completed the National Levee Inventory for both federal and nonfederal levees, using a levee hazard potential classification system	Improve protection for human lives and infrastructure that are protected by levees and provide consistent protection
Inspect and bring up to code, all existing levee systems using updated hydrology and hydraulic analyses that incorporate the impacts of urbanization and climate change	Improve protection for human lives and infrastructure that are protected by levees
Increase or reallocate reservoir storage	Assure that there is enough reservoir storage space both for water storage and for flood protection
Develop flexible policies, at all levels of government, for upgrading dams and levees to handle continually changing flood risks	Provide consistent flood protection as the flood risks change
Develop a portfolio of nonstructural measures to manage flood risk and reduce the reliance on dams, levees, and flood barriers	Reduce the reliance on infrastructure for flood protection
Ensure that operation and maintenance plans cover all aspects of a complex levee system	Improve protection for human lives and infrastructure that are protected by levees
Require insurance where appropriate and create emergency action plans for levee-protected areas	Increase awareness of the potential for loss from flood and allow for better preparation for flood events
Mechanism for coordinating dam safety regulations across all levels of government for all NID dams with high-hazard potential	Improve consistency of dam safety
Perform detailed structural assessments of high-hazard dams and spillways, and retrofit as needed	Mitigate the risk caused by our highest-hazard dams
Upgrade and maintain all types of in-channel and off-channel flood diversion structures	Make sure infrastructure is available and functional to divert floodwaters into areas where they do less harm

(Continued)

Table 3-8. *Potential Adaptation Actions for Dams and Reservoirs, Levees, and Flood Barriers. (Continued)*

Potential adaptation actions	How they will help?
Routine sediment monitoring/ assessment of reservoirs and prioritize reservoirs for sediment management	Increase awareness of the loss of flood pool capacity caused by sedimentation
Improve weather and streamflow forecasting (subseasonal to seasonal)	Allow for better preparation for flood events and movement of water in anticipation of flood events as needed to minimize risk
Perform routine inspection and maintenance of all flood risk management structures and ensure sufficient funding for this activity at all levels of government	Avoid costly surprises
Ask the Association of Dam Safety Officials to convene a working group to address the need for systematic inspection and ensure that maintenance issues are addressed for private dams, levee systems, and diversions, with priority on high-hazard structures	Ensure consistent safety between private and public infrastructure
Develop alternate water supply storage (surface water and groundwater) and flood protection infrastructure with different failure modes	Mitigate decreasing supplies because of increasing temperatures/aridification, longer and more intense droughts, and decreasing ability to store water in upland forests
Perform thinning and prescribe burns in overgrown upland forests to prevent catastrophic wildfire. Reinitiate healthy fire into forest ecosystems	Reduce sedimentation and water quality impacts from wildfire on reservoirs. Reduce the threat of sedimentation/debris flow preventing safe operation of dams and diversions
In all future infrastructure repairs or upgrades, consider the potential failure mechanism of the system component being repaired or replaced (both in today's climate and for likely future climate stressors) and build in redundancy using system components with other failure modes, to build resilience in the overall system	Build in the engineering today to avoid preventable future failures

(Continued)

Table 3-8. Potential Adaptation Actions for Dams and Reservoirs, Levees, and Flood Barriers. (Continued)

Potential adaptation actions	How they will help?
Create the funding streams needed to support building in redundancy to address future climate change and create the regulatory environment around that funding to assure it is applied as intended	Assure we do not miss opportunities to increase our adaptive capacity because of a lack of funding and awareness
Assure that the legal system allows for innovation	Save money in the long term
	Provide the flexibility needed to adapt operations and infrastructure to changing conditions and demands

showed that actions at particular locations can have unpleasant impacts on others. For example, Holleman and Stacey (2014) demonstrated that shoreline protection within South San Francisco Bay would lead to an increase of up to 0.2 m (0.66 ft) in the northern bay water levels because of the interaction between tidal dynamics and shorelines.

Despite its great importance, only few research studies have been devoted to investigating the effect of levee failure on critical infrastructure. Wang et al. (2018) showed that measures to prevent flooding along a shoreline in one location may increase inundation elsewhere in the system. Peng and Song (2018) assessed the cost efficiency and proposed corresponding benefit improvements by estimating the avoided damages because of the implementation of levee projects in Miami. Papakonstantinou (2019) used a simulation-based optimization model to study how delayed travel time caused by various projected flood scenarios is affected by the construction of levees at different locations. However, there is still great need in this field for assessing the imposed damage to infrastructure because of different modes of levee failure caused by realistic flooding projections owing to climate change.

Figure 3-4 also illustrates the relationships of both the dams and reservoirs and flood protection sectors and the transportation, transit, and aviation, including rail, and navigation/ports and harbor sectors. Both sectors rely on an effective network of roads and bridges as well as transit to provide capabilities for regular inspection and maintenance and to ensure efficient evacuation should there be a failure. Full functioning within the dams and reservoirs and flood protection sectors also relies on the system of essential nodes and links that the transit and aviation sector provides within the labor, energy, and/or material networks. The navigation sector, on the contrary, relies on the dams and reservoirs sector in the matter of water allocation to ensure that sufficient water levels are maintained within navigation channels. A codependence also exists with transit

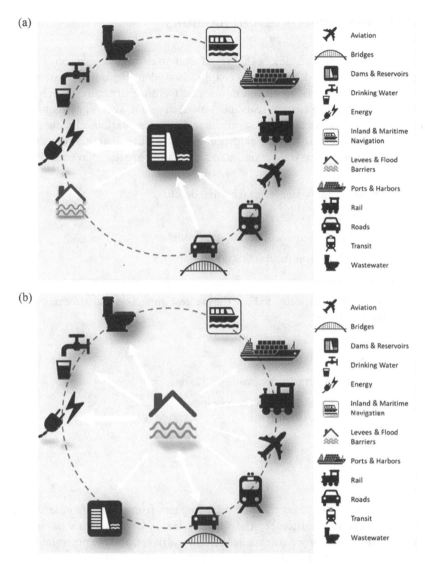

Figure 3-4. Sectoral interdependencies involving sectors: (a) dams and reservoirs, (b) levees and other flood barriers. The arrow thickness indicates the level of criticality of sector interdependence. The double arrows indicate the dependence of both sectors on each other.

and aviation and ports and harbors because of the fact both sectors depend on the reliable service of levees and other flood barriers, whether it be inland or coastal, for protection in the case of a sudden extreme precipitation and/or wind/wave event or, specifically pertaining to ports and harbors, because of the reality of a gradual rise in sea levels over time.

3.5 NAVIGATION/PORTS AND HARBORS

Navigation, which includes both inland and maritime navigation, may be significantly impacted by climate change. Inland navigation infrastructure includes river channels, locks and dams, and training structures for channels such as dikes. Port infrastructure includes piers, channels, anchorages, equipment for loading and unloading ships, and support systems. Various climate change drivers or stressors can affect navigation infrastructure. Potential drivers that could have a significant impact on inland navigation are the following (PIANC 2010):

- Precipitation and water levels (channel depths),
- Water temperature,
- Frequency and intensity of floods and droughts,
- Modification to river morphology, and
- Ice and snow cover,

Drivers that are expected to have the greatest impact on maritime navigation include the following:

- Sea level,
- Wind speed,
- Tide and storm surge propagation and range,
- Coastal hydrodynamics and morphology,
- Storm events,
- Sea chemistry,
- Ice cover, and
- Icing.

This section discusses the potential impacts from each of these drivers on navigation infrastructure, the current status of America's navigation infrastructure regarding resilience against these drivers, and the potential future responses to increasing resiliency.

3.5.1 Impacts

3.5.1.1 Inland Navigation

Changes in the amount and form of precipitation, both seasonally and overall, can impact the channel depths and velocity and affect sedimentation processes such as bank failure, scour, accretion, and erosion. Changes in channel depths and the movement of sedimentation would require changes in the amount of maintenance dredging that is required to maintain minimum channel depths for navigation. Navigation structures may also be affected through changes in stability and operational efficiency because of different loading distributions.

Changes in precipitation could lead to a change in the frequency and intensity of floods and droughts. Floods are a hazard to navigation. High flows increase the velocity of the water in the channel. Towboats and full barge loads have more difficulty maneuvering, leading to an increased risk of boat accidents. Major floods may cause an inland waterway to close to navigation. Flooding can also lead to more shoaling in the channel. Drought also affects barge traffic. As channel depths decrease because of drought, the draft of barges will be restricted, causing an increase in barge operating costs. The channel may eventually close to barge traffic if channel depths decrease too much.

An increase or decrease in sediment loading will have significant impacts on river morphology in the form of river bed erosion, dune development, and floodplain accretion, which would require adaptation measures such as additional dredging. Such changes in sedimentation may also affect the functioning of vital ecosystems.

An increase in water temperature because of warmer air temperatures would result in an increased frequency of oxygen deficits given similar nutrient loadings, which will affect navigation through regulations that offer additional protection to enhance and preserve vital riverine and estuarine ecosystems.

Changes in the magnitude and variability of water temperature may also result in shorter periods of ice cover and can affect its thickness and roughness. Sudden increases or decreases in temperature at either end of the winter season can result in a higher frequency of ice jams, even in locations where are ice jams are not expected to occur. Earlier breakup of ice can also result in increased ice strength should temperatures fall back below freezing.

3.5.1.2 Maritime Navigation

Changes in sea level will not necessarily have negative impacts on navigation itself but would affect infrastructure within ports and harbors. Higher levels would enhance coastal erosion because of increased penetration of wave energy along the coastline and into harbors, increase salinity within bays and estuaries, cause more frequent overtopping and lowland flooding, create reduced clearance between vessels and bridges, increase the level at which wave forces impact a building, alter patterns of sediment movement and deposition, and increase the amount of pumping required to maintain dry-docks dewatered, among other effects. In contrast, potential positive impacts include a reduction in the need for dredging and greater under keel clearance for vessels.

An increase in wind speed has the potential to increase wave amplitude, create a need to alter preferred shipping routes because of curved or narrow shipping channels being more difficult to maneuver, increase the downtime of wind-sensitive vessels, and require larger areas dedicated to the anchoring of vessels because of reduced time windows for calm weather at high-risk terminals and increased berthing time and delayed departures of ships at terminals.

Any response to changes in coastal hydrodynamics or morphology would vary based on the location and may include the opening, closing, narrowing, or widening of navigation channels, changes in maintenance dredging requirements, and the erosion of beaches protecting port infrastructure because of changes in current velocities.

Changes in storm events refer to changes in storm duration, frequency, and track, which may cause reduced accessibility of ports, increased downtime, and the need for additional capacity at cargo terminals during such events. A higher frequency of intense storms may also cause permanent erosion of the coast, which would lead to a reduction in viable land and potential structure degradation. Other impacts may include reduced visibility and reductions in available solar energy.

Potential changes in sea chemistry resulting from climate change include an increase in salt concentrations, which may lead to an increase in corrosion and deterioration of vessels and port infrastructure.

The duration of periods of ice cover within waterways is expected to decrease because of warmer temperatures. Such conditions would lead to improved accessibility of waterways seasonally or, in some cases, permanently. Less ice cover would result in more freshwater in rivers and potentially cause more ice to form in estuaries, altering the salinity and chemistry of these water bodies.

Icing on vessels occurs because of a combination of cold temperatures and high wind speeds, which, on contact, results in the immediate freezing of spray blown off the sea. Substantial icing can build up to a weight that is sufficient to raise the center of gravity and lower the freeboard of vessels, which cause reduced stability and can eventually lead to vessels capsizing. Changes in icing conditions will also affect the stability of structures along the coast.

Substantial amounts of excess carbon dioxide have been absorbed by the oceans, which has increased ocean water acidity. Using the highest emission scenario (RCP8.5), ocean surface acidity is projected to increase by 100% to 150% compared with late twentieth-century levels by the year 2100 (USGCRP 2018).

3.5.2 Status

3.5.2.1 Inland Navigation

Among the climate stressors previously summarized, the most concerning for the navigation sector are forecasted changes in river flows and channel depths because of changing climate. The system performance during observed droughts and floods can show the current vulnerability of the inland navigation system. The system of locks and dams on the waterways supports channel depths during low flow periods, but in some reaches, there are no locks and dams to maintain navigation depths. These reaches are particularly vulnerable to low flows. The Middle Mississippi River, the segment of the Mississippi River from its confluence with the Missouri River at St. Louis, Missouri, to its confluence with the Ohio River at Cairo, Illinois, is one such reach. The experiences of the 1988 drought and more recent less severe droughts provide an illustration of its vulnerability. USACE increased dredging in the affected channels. The Coast Guard implemented restrictions on the size, draft, and configuration of tows operating when channel depths decrease. Barge shipping rates increase when there are channel restrictions because tow sizes and barge capacities are reduced.

Shippers do have options when the channels are restricted or closed because of drought. Shipments can be sent by alternate modes of transportation, particularly rail. For example, during the 1988 drought, the rail line from Chicago to New Orleans had a cost advantage and greatly increased the amount it carried (Riebsame et al. 1991). Midwestern shippers can also shift to export from ports to the Pacific Northwest or through the Great Lakes. This flexibility provides some resilience to the overall transportation system.

The 1993 flood on the Mississippi River caused the channel to be closed for barge traffic. The Middle Mississippi River was closed for more than 6 weeks in July and August 1993. There were limitations on barge traffic for the next 3 months. The flood also affected bridges and rail and roads in floodplains, causing disruptions in the overall transportation system (Changnon 1996).

Some studies have projected that droughts may become more severe in the future. For example, one study projects lower Ohio River main stem flows during the fall season, which may be 25% to 35% lower during the period 2070 to 2099 compared with the average flows during the period 1952 to 2011. The same study projects that the Ohio River main stem flow during spring may increase by as much as 25% to 35% by the end of the twenty-first century when compared with the average flows during 1952 to 2011 (Drum et al. 2017).

3.5.2.2 Ports

The recent impact of Superstorm Sandy on the port of New York and New Jersey and that of Hurricane Harvey on Houston provide examples of the current vulnerability of ports to coastal storms. Sandy produced a storm surge of more than 14 ft in New York Harbor. The entire port was shut for 1 week, whereas parts of the harbor were shut for longer periods because of shoaling and the need to check navigation aids. Port electrical equipment, power substations, motors on gantry cranes, and pumping stations were submerged by saltwater. Many loaded containers were also inundated. Approximately 25,000 shipping containers were diverted to other ports, thus increasing the costs for shippers. The primary impact from Harvey was because of the unprecedented rainfall, which brought increased sediment to the Houston Ship Channel. The sediment caused reduced channel depths, resulting in higher shipper costs because ships had to reduce drafts. Increased dredging was needed to restore channel depths. The impact from Sandy and Harvey was short term as the ports recovered from the damage, but there were major economic costs to shippers for diverting or reducing shipments and to port authorities for repair and recovery of their facilities.

3.5.3 Responses

3.5.3.1 Inland Navigation

Increased frequency and magnitude of droughts would increase the number of channel closures and restrictions and reduce the efficiency of the inland navigation industry. The system is vulnerable to a severe multiyear drought that closes the channel, because there may not be enough rail capacity that could make up for the loss of barge capacity. Even though the system of locks and dams is not the most vulnerable reach of the system, reduced river flow could affect the filling of navigation locks. Water-reducing chambers could promote increased efficiency of navigation locks. One potential adaptation measure includes better communication systems, particularly with regard to gauge data and channel hazards. Monitoring of incoming high flows from tributaries, bridge clearances, and channel velocities near locks and dams could improve navigation safety.

3.5.3.2 Ports

In southeast coastal Virginia, sea levels have risen about 1.5 ft in the last 100 years. As part of port renovation, the port of Virginia is raising electric power stations several feet off the ground, and data servers are being moved as far from the ocean as possible (WSJ 2019). Although the supporting infrastructure may be vulnerable to flooding, piers are in general built as high as 14 ft above high tide and, therefore, are less vulnerable to flooding in the near future.

3.5.4 Interdependencies

Navigation can be an element of a managed water resources system with multiple objectives. Other potential competing water uses include municipal and industrial water supply, irrigation, energy production including hydropower and cooling for thermal power plants, ecosystem flow needs, and recreation. In a reservoir, conservation storage to support these uses must be balanced with flood storage space. Balancing water supply with various other water needs (e.g., irrigation, hydropower, and domestic consumption) may become more complicated as a greater occurrence of drought will exacerbate any shortfalls between supply and demand for each of these sectors, thus reducing the amount of water available for navigation and port purposes. Potential increased flood risk may limit the feasibility of reallocating flood storage space to conservation storage. For these reasons, the dams and reservoirs and flood protection sectors are identified as key interdependencies in Figure 3-5.

Inland navigation is also a part of the transportation network and competes with trucks and rail for the shipment of cargoes. During droughts, higher rates for barge shipments may cause shippers to shift to rail or possibly trucks. A river closure because of drought would make land transportation the only available option. During floods, barge, rail, and trucks may all be affected. Bridges and roads and rail in floodplains may be closed. Coastal ports are also interconnected with the trucking and rail systems. A major coastal storm can force closures of major road and rail routes, which may force a coastal port to close even if its infrastructure is not damaged. For these reasons, the roads and bridges and transit sectors are indicated as secondary interdependencies, as shown in Figure 3-5.

A functioning port also requires energy to support operations, a labor force that can get to work via effective road and transit networks, and effective drinking water and sanitation services to support the labor force. Energy, transportation, housing, and other supporting infrastructure could all be disrupted during a

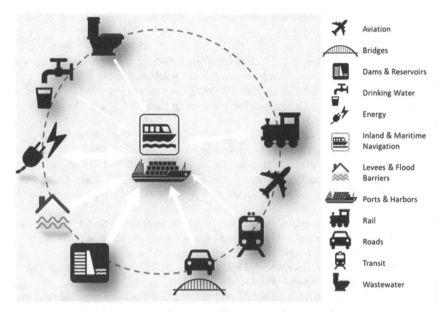

Figure 3-5. *Sectoral interdependencies involving the navigation sector, which includes inland and maritime navigation as well as ports and harbors. The arrow thickness indicates the level of criticality of sector interdependence. The double arrow indicates the dependence of both sectors on each other.*

major coastal storm. In addition, the energy sector relies on the transportation, transit, and navigation sectors for the transportation of energy products essential to its continuous operation and maintenance; any disruption in any one of these transport networks could have drastic impacts on the energy and, as a result, the ports and harbors sectors.

References

American Rivers. 2014. "72 Dams removed to restore rivers in 2014." Press Release. Accessed July 23, 2019. https://www.circleofblue.org/wp-content/uploads/2015/01/AmericanRivers_Dam-List-2014.pdf.

APTA (American Public Transit Association). 1994. "Glossary of transit terminology." Washington, DC: APTA Governing Boards Committee. Accessed November 1, 2018. https://www.apta.com/resources/reportsandpublications/Documents/Transit_Glossary_1994.pdf.

ASCE. 2017. *Infrastructure report card. A comprehensive assessment of America's infrastructure.* Reston, VA: ASCE.

ASCE CACC (Committee on Adaptation to a Changing Climate). 2015. *Adapting infrastructure and civil engineering practice to a changing climate,* edited by J. R. Olsen. Reston, VA: ASCE.

ASCE CACC. 2018. *Climate-resilient infrastructure: Adaptive design and risk management,* edited by B. M. Ayyub. ASCE Manuals and Reports on Engineering Practice No. 140. Reston, VA: ASCE.

ASDSO (Association of State Dam Safety Officials). 2019a. *State performance and current issues.* Lexington, KY: ASDSO.

ASDSO. 2019b. *Dam failures and incidents.* Lexington, KY: ASDSO.

Bates, M. E., and J. R. Lund. 2013. "Delta subsidence reversal, levee failure, and aquatic habitat—A cautionary tale." *San Francisco Estuary Watershed Sci.* 11 (1): 1–20.

Behr, J. G., R. Diaz, and M. Mitchell. 2016. "Building resiliency in response to sea level rise and recurrent flooding: Comprehensive planning in Hampton Roads." *Virginia News Lett.* 92 (1): 1–6.

Bennett, V., X. Lv, M. Zeghal, T. Abdoun, B. Yazici, and A. Marr. 2014. "Multi-scale monitoring for health assessment of levees in New Orleans." In *Proc., Geo-Congress 2014: Geo-Characterization and Modeling for Sustainability,* 252–261. Reston, VA: ASCE.

Bertin, X., K. Li, A. Roland, Y. J. Zhang, J. F. Breilh, and E. Chaumillon. 2014. "A modeling-based analysis of the flooding associated with Xynthia, central Bay of Biscay." *Coast. Eng.* 94: 80–89.

Biello, D. 2013. "What role does climate change play in tornadoes?" *Scientific American,* May 21, 2013. Accessed January 15, 2019. https://www.scientificamerican.com/article/kevin-trenberth-on-climate-change-and-tornadoes/.

Bierkandt, R., M. Auffhammer, and A. Levermann. 2015. "US power plant sites at risk of future sea-level rise." *Environ. Res. Lett.* 10 (12): 124022.

Billington, D. P., D. C. Jackson, and M. V. Melosi. 2005. *The history of large federal dams: Planning, design, and construction in the era of big dams.* Denver: Dept. of the Interior, Bureau of Reclamation.

Bles, T., J. Bessembinder, M. Chevreuil, P. Danielsson, S. Falemo, A. Venmans, et al. 2016. "Climate change risk assessments and adaptation for roads—Results of the ROADAPT project." *Transp. Res. Procedia* 14: 58–67.

Bles, T., Y. Ennesser, J.-J. Fadheuilhe, S. Falemo, B. Lind, M. Mens, et al. 2010. *Risk management for roads in a changing climate. A guidebook to the RIMAROCC method.* ERA-NET Road.

Brooks, B. A., G. Bawden, D. Manjunath, C. Werner, N. Knowles, J. Foster, et al. 2012. "Contemporaneous subsidence and levee overtopping potential, Sacramento-San Joaquin Delta, California." *San Francisco Estuary Watershed Sci.* 10 (1).

Brozović, N., D. L. Sunding, and D. Zilberman. 2007. "Estimating business and residential water supply interruption losses from catastrophic events." *Water Resour. Res.* 43 (8): W08423.

Chang, S. E. 2016. Vol. 1 of *Socioeconomic impacts of infrastructure disruptions.* Oxford, UK: Oxford University Press.

Changnon, S. A. 1996. "Impacts on transportation systems: Stalled barges, blocked railroads, and closed highways." In Chap. 8 in *The great flood of 1993: Causes, impacts, and responses,* edited by S. A. Changnon, 183–204. Boulder, CO: Westview Press.

Clark, M. P., R. L. Wilby, E. D. Gutmann, J. A. Vano, S. Gangopadhyay, A. W. Wood, et al. 2016. "Characterizing uncertainty of the hydrologic impacts of climate change." *Clim. Change Rep.* 2 (2): 55–64.

Clarke, G. R. T., D. A. B. Hughes, S. L. Barbour, and V. Sivakumar. 2006. "The implications of predicted climate changes on the stability of highway geotechnical infrastructure: A case study of field monitoring of pore water response." In *EIC Climate Change Technology,* 1–10. New York: IEEE.

Conant, R. T., M. G. Ryan, G. I. Ågren, H. E. Birge, E. A. Davidson, P. E. Eliasson, et al. 2011. "Temperature and soil organic matter decomposition rates—Synthesis of current knowledge and a way forward." *Global Change Biol.* 17 (11): 3392–3404.

Davidson, E. A., and I. A. Janssens. 2006. "Temperature sensitivity of soil carbon decomposition and feedbacks to climate change." *Nature* 440 (7081): 165–173.

de Bruijn, K., J. Buurman, M. Mens, R. Dahm, and F. Klijn. 2017. "Resilience in practice: Five principles to enable societies to cope with extreme weather events." *Environ. Sci. Policy* 70: 21–30.

Dehn, M., G. Bürger, J. Buma, and P. Gasparetto. 2000. "Impact of climate change on slope stability using expanded downscaling." *Eng. Geol.* 55 (3): 193–204.

De Neufville, R., and S. Scholtes. 2011. *Flexibility in engineering design.* Cambridge, MA: MIT Press.

Dettinger, M. D. 2013. "Atmospheric rivers as drought busters on the U.S. west coast." *J. Hydrometeorol.* 14 (6): 1721–1732.

DOE (US Department of Energy). 2015. *QER report: Energy transmission, storage, and distribution infrastructure.* Washington, DC: DOE.

Drum, R. G., J. Noel, J. Kovatch, L. Yeghiazarian, H. Stone, J. Stark, et al. 2017. *Ohio River Basin—Formulating climate change mitigation/adaptation strategies through regional collaboration with the ORB alliance.* Civil Works Technical Rep. CWTS 2017-01. Alexandria, VA: US Army Corps of Engineers, Institute for Water Resources.

Du, H., L. V. Alexander, M. G. Donat, T. Lippmann, A. Srivastava, J. Salinger, et al. 2019. "Precipitation from persistent extremes is increasing in most regions and globally." *Geophys. Res. Lett.* 46 (11): 6041–6049.

Dunbar, J., J. Llopis, G. Sills, E. Smith, R. Miller, R. Ivanov, et al. 2007. *Condition assessment of levees, U.S. section of the international boundary and water commission.* Technical Rep. No. TR-03-4. Vicksburg, MS: US Army Engineer Geotechnical and Structures Laboratory.

Dupray, S., R. Tourment, R. Pohl, H. Schelfhout, T. Williamson, K. Gamst, et al. 2010. *International levee handbook—Scoping report.* London: Construction Industry Research and Information Association.

Dyer, M. R., S. Utili, and M. Zielinski. 2009. "Field survey of desiccation fissuring of flood embankments." *Water Manage.* 162 (3): 221–232.

EIA (US Energy Information Administration). 2017. "Table 1.1A: Net generation from renewable sources." *Electric Power Monthly.* Washington, DC: EIA. Accessed February 15, 2019. https://www.eia.gov/electricity/monthly/epm_table_grapher.php?t=epmt_1_01_a.

EIA. 2018. *Electric power monthly.* Washington, DC: EIA.EPA. 2014. *Flood Resilience: A Basic Guide for Water and Wastewater Utilities.* EPA-817-B-14-006. Washington, DC: EPA.

EPA. 2017. *Climate impacts on energy.* Washington, DC: EPA.

EOP (Executive Office of the President). 2013. *Preparing the United States for the impacts of climate change.* Executive Order 13653. Washington, DC: EOP.

FEMA. 2004a. *Federal guidelines for dam safety.* FEMA 93. Washington, DC: FEMA.

FEMA. 2004b. *Federal guidelines for dam safety: Hazard potential classification system for dams.* Washington, DC: FEMA.

FEMA. 2013. *Federal guidelines for dam safety: Emergency action planning for dams.* FEMA 64. Washington, DC: FEMA.

FEMA. 2015. *FEMA national dam safety program fact sheet.* FEMA P-1069. Washington, DC: FEMA.

FHWA (Federal Highway Administration). 2018. *2020/2024 Pilot program: Resilience and durability to extreme weather.* Washington, DC: FHWA.

Fimrite, P., and J. Palomino. 2018. "60,000 without power as PG&E shuts down lines over wildfire fears." *SFGate,* Accessed October 15, 2018. https://www.sfgate.com/california-wildfires/article/PG-E-warns-it-may-shut-off-power-amid-red-flag-13306256.php.

FTA (Federal Transit Administration). 2017. *National transit database, glossary.* Washington, DC: Office of Budget and Policy, Federal Transit Administration, US Dept. of Transportation.

Gariano, S. L., and F. Guzzetti. 2016. "Landslides in a changing climate." *Earth Sci. Rev.* 162: 227–252.

Gibbs, A. E., and B. M. Richmond. 2016. *National assessment of shoreline change— Historical shoreline change along the north coast of Alaska, U.S.–Canadian border to Icy Cape.* Open-File Rep. 2015-1048. Washington, DC: US Geological Survey.

Gimon, E., M. O'Boyle, C. Clack, and S. Mckee. 2019. "The coal cost crossover: Economic viability of existing coal compared to new local wind and solar resources." *Energy Innovation and Vibrant Clean Energy,* Accessed March 24, 2019. https:// energyinnovation.org/publication/the-coal-cost-crossover/.

Government of Puerto Rico. 2018. "Transformation and innovation in the wake of devastation: An economic and disaster recovery plan for Puerto Rico." Accessed December 6, 2018. http://www.p3.pr.gov/assets/pr-draft-recovery-plan-for-comment-july-9-2018.pdf.

Grady, B. 2014. "Sea level rise threatens Oakland's sewer system." *Climate Central.* Accessed June 17, 2014. https://www.climatecentral.org/news/sea-level-rise-oakland-sewer-17567.

Gulliver, J. S., and R. E. A. Arndt. 1991. *Hydropower engineering handbook.* New York: McGraw-Hill.

Holleman, R. C., and M. T. Stacey. 2014. "Coupling of Sea Level Rise, Tidal Amplification, and Inundation." *J. Phys. Oceanogr.* 44 (5): 1439–1455.

Hungr, O., J. Corominas, and E. Eberhardt. 2005. "Estimating landslide motion mechanisms, travel distance and velocity." In *Landslide risk management,* edited by O. Hungr, R. Fell, R. Couture, and E. Eberhardt, 99–128. London: Taylor and Francis.

IAWG (Immediate Action Workgroup). 2009. "Recommendations Report to the Governor's Subcabinet on Climate Change". Juneau, AK: Alaska SubCabinet on Climate Change, Immediate Action Workgroup.

Iverson, R. M., M. E. Reid, M. Logan, R. G. LaHusen, J. W. Godt, and J. P. Griswold. 2011. "Positive feedback and momentum growth during debris-flow entrainment of wet bed sediment." *Nat. Geosci.* 4 (2): 116–121.

Jacobs, J. M., L. R. Cattaneo, W. Sweet, and T. Mansfield. 2018b. "Recent and future outlooks for nuisance flooding impacts on roadways on the US East Coast." *Transp. Res. Rec.* 2672 (2): 036119811875636.

Jacobs, J. M., M. Culp, L. Cattaneo, P. Chinowsky, A. Choate, S. DesRoches, et al. 2018a. "Transportation." In Chap. 12 in Vol. 2 of *Impacts, risks, and adaptation in the United States: Fourth National Climate Assessment,* edited by D. R. Reidmiller, C. W. Avery, D. R. Easterling, K. E. Kunkel, K. L. M. Lewis, T. K. Maycock, et al., 479–511. Washington, DC: US Global Change Research Program.

Jasim, F. H., F. Vahedifard, A. Alborzi, H. Moftakhari, and A. Aghakouchak. 2020. "Effect of compound flooding on performance of earthen levees." In *Geo-Congress 2020: Engineering, Monitoring, and Management of Geotechnical Infrastructure,* Geotechnical Special Publication 316, edited by J. P. Hambleton, R. Makhnenko, and A. S. Budge, 707–716. Minneapolis: ASCE.

Jasim, F. H., F. Vahedifard, E. Ragno, A. AghaKouchak, and G. Ellithy. 2017. "Effects of climate change on fragility curves of earthen levees subjected to extreme precipitations." In *Geo-Risk 2017: Geotechnical Risk from Theory to Practice,* Geotechnical Special Publication No. 285, edited by J. Huang, G. A. Fenton, L. Zhang, and D. V. Griffiths, 498–507, Denver: ASCE.

Jones, J. A. A. 2010. "Soil piping and catchment response." *Hydrol. Processes* 24 (12): 1548–1566.

Karnauskas, K. B., J. K. Lundquist, and L. Zhang. 2018. "Southward shift of the global wind energy resource under high carbon dioxide emissions." *Nat. Geosci.* 11 (1): 38–43.

Kim, Y., D. A. Eisenberg, E. N. Bondank, M. V. Chester, G. Mascaro, and B. S. Underwood. 2017. "Fail-safe and safe-to-fail adaptation: Decision-making for urban flooding under climate change." *Clim. Change* 145 (3–4): 397–412.

Knight, B., and T. Rummel. 2014. "DamNation". Patagonia. Accessed May 10, 2019. http://damnationfilm.com.

Kovacevic, K., D. M. Potts, and P. R. Vaughan. 2001. "Progressive failure in clay embankments due to seasonal climate changes." In Vol. 3 of *Proc., Int. Conf. Soil Mechanics and Geotechnical Engineering*, 2127–2130. Lisse [u.a.]: Balkema.

Larsen, K., J. Larsen, M. Delgado, W. Herndon, and S. Mohan. 2017. *Assessing the effect of rising temperatures: The cost of climate change to the US power sector.* New York: Rhodium Group.

Lazard. 2017. "Lazard's levelized cost of energy analysis. Version 11.0." Accessed February 17, 2019. https://www.lazard.com/perspective/levelized-cost-of-energy-2017/.

Lee, E. M., and D. K. C. Jones. 2004. *Landslide risk assess.* London: Thomas Telford.

Lee, S., M. Li, and F. Zhang. 2017. "Impact of sea-level rise on tidal range in Chesapeake and Delaware Bays." *J. Geophys. Res.: Oceans* 122 (5): 3917–3938.

Leshchinsky, B., F. Vahedifard, H.-B. Koo, and S.-H. Kim. 2015. "Yumokjeong landslide: An investigation of progressive failure of a hillslope using the finite element method." *Landslides* 12 (5): 997–1005.

Li, Z., and H. Fang. 2016. "Impacts of climate change on water erosion: A review." *Earth-Sci. Rev.* 163: 94–117.

Lopez-Cantu, T., and C. Samaras. 2018. "Temporal and spatial evaluation of stormwater engineering standards reveals risks and priorities across the United States." *Environ. Res. Lett.* 13 (7): 074006.

Lu, N., and W. J. Likos. 2004. *Unsaturated soil mechanics.* Hoboken, NJ: Wiley.

Malmqvist, B., and S. Rundle. 2002. "Threats to the running water ecosystems of the world." *Environ. Conserv.* 29 (2): 134–153.

Marsh, J. 2019. "How hot do solar panels get? Effect of temperature on solar performance." *EnergySage.* Accessed February 10, 2019. https://news.energysage.com/solar-panel-temperature-overheating/.

Mattsson, L.-G., and E. Jenelius. 2015. "Vulnerability and resilience of transport systems—A discussion of recent research." *Transp. Res. Part A: Policy Pract.* 81: 16–34.

Maxwell, J. T., P. A. Knapp, J. T. Ortegren, D. L. Ficklin, and P. T. Soulé. 2017. "Changes in the mechanisms causing rapid drought cessation in the southeastern United States." *Geophys. Res. Lett.* 44 (24): 12476–12483.

McCully, P. 2001. *Silenced rivers: The ecology and politics of large dams.* Ann Arbor, MI: Univ. of Michigan Press.

Moftakhari, H. R., A. AghaKouchak, B. F. Sanders, M. Allaire, and R. A. Matthew. 2018. "What is nuisance flooding? Defining and monitoring an emerging challenge." *Water Resour. Res.* 54 (7): 4218–4227.

Moftakhari, H. R., A. AghaKouchak, B. F. Sanders, and R. A. Matthew. 2017. "Cumulative hazard: The case of nuisance flooding." *Earth's Future* 5 (2): 214–223.

Mori, N., T. Takahashi, and The 2011 Tohoku Earthquake Tsunami Joint Survey Group. 2012. "Nationwide post event survey and analysis of the 2011 Tohoku earthquake tsunami." *Coast. Eng.* 54 (1): 1250001-1–1250001-27.

Mount, J., and R. Twiss. 2005. "Subsidence, sea level rise, and seismicity in the Sacramento–San Joaquin Delta." *San Francisco Estuary Watershed Sci.* 3 (1): 5.

NASEM (National Academies of Sciences, Engineering, and Medicine). 2014. *Response to extreme weather impacts on transportation systems.* Washington, DC: National Academies Press.

NASEM. 2016. *Transportation resilience: Adaptation to climate change.* Washington, DC: National Academies Press.

NGC (Natural Gas Council). 2017. "Natural gas systems: Reliable & resilient." Accessed February 15, 2019. https://www.aga.org/sites/default/files/ngc_reliable_resilient_nat_gas_white_paper.pdf.

Nelson, L. 2018. "Resilience and continuity in an interconnected and changing world. London, UK: Bank of England." In *Speech given at the 20th Annual Operational Risk Europe Conf.* Accessed June 13, 2018. https://www.bankofengland.co.uk/-/media/boe/files/speech/2018/resilience-and-continuity-in-an-interconnected-and-changing-world-speech-by-lyndon-nelson.pdf?la=en&hash=3DFD465DE00C545FA39255384DD660AF7B0C8A9F.

NYSDOT (New York State DOT). 2011. *Mainstreaming climate change adaptation strategies into New York State Department of Transportation's Operations.* New York: NYSDOT.

Nilsson, C., and K. Berggren. 2000. "Alterations of Riparian Ecosystems Caused by River Regulation: Dam operations have caused global-scale ecological changes in riparian ecosystems. How to protect river environments and human needs of rivers remains one of the most important questions of our time." *BioScience* 50 (9): 783–792.

NRC (National Research Council). 2012. *Dam and levee safety and community resilience: A vision for future practice.* Washington, DC: National Academies Press.

Nyambayo, V. P., D. M. Potts, and T. I. Addenbrooke. 2004. "The influence of permeability on the stability of embankments experiencing seasonal cyclic pore water pressure changes." *Adv. Geotech. Eng.* 2: 898–910.

ORR (NYC Mayor's Office of Recovery and Resiliency). 2019. "Climate resiliency design guidelines (Version 3.0)." Accessed November 1, 2018. https://www1.nyc.gov/assets/orr/pdf/NYC_Climate_Resiliency_Design_Guidelines_v3-0.pdf.

Papakonstantinou, I., J. Lee, and S. M. Madanat. 2019. "Optimal levee installation planning for highway infrastructure protection against sea level rise." *Transp. Res. D Transp. Environ.* 77: 378–389.

Pendergrass, A. G., and R. Knutti. 2018. "The uneven nature of daily precipitation and its change." *Geophys. Res. Lett.* 45 (21): 11980–11988.

Peng, B., and J. Song. 2018. "A Case Study of Preliminary Cost-Benefit Analysis of Building Levees to Mitigate the Joint Effects of Sea Level Rise and Storm Surge." *Water* 10 (2): 169.

PIANC (World Association for Waterborne Transport Infrastructure). 2010. *Waterborne transport, ports and waterways: A review of climate change drivers, impacts, responses and mitigation.* Report of PIANC EnviCom Task Group 3. Brussels, Belgium: PIANC.

Pohl, M. M. 2002. "Bringing down our dams: Trends in American dam removal rationales." *J. Am. Water Resour. Assoc.* 38 (6): 1511–1519.

Port, P. S., and S. A. Hoover. 2011. "Anticipating California levee failure: The state of the delta levees and government preparation and response strategies for protecting natural resources from freshwater oil spills." In *Proc., Int. Oil Spill Conf., 1, Portland, OR, USA: IOSC (International Oil Spill Conference)*, http://iosc.org/.

Potts, D. M., K. Kovacevic, and P. R. Vaughan. 1997. "Delayed collapse of cut slopes in stiff clay." *Géotechnique* 47 (5): 953–982.

Prein, A. F., C. Liu, K. Ikeda, R. Bullock, R. M. Rasmussen, G. J. Holland, et al. 2020. "Simulating North American mesoscale convective systems with a convection-permitting climate model." *Clim. Dyn.* 55: 95–110.

Ragno, E., A. AghaKouchak, C. A. Love, L. Cheng, F. Vahedifard, and C. H. R. Lima. 2018. "Quantifying changes in future intensity–duration–frequency curves using multimodel ensemble simulations." *Water Resour. Res.* 54 (3): 1751–1764.

Reisner, M. 1993. *Cadillac desert: The American west and its disappearing water.* Revised edition. New York: Penguin Books.

Renwick, W. H., S. V. Smith, J. D. Bartley, and R. W. Buddemeier. 2005. "The role of impoundments in the sediment budget of the conterminous United States." *Geomorphology* 71 (1–2): 99–111.

Riebsame, W. E., S. A. Changnon, and T. R. Karl. 1991. *Drought and natural resources management in the United States impacts and implications of the 1987–89 drought.* Boulder, CO: Westview Press.

Robinson, J. D., and F. Vahedifard. 2016. "Weakening mechanisms imposed on California's levees under multiyear extreme drought." *Clim. Change* 137 (1–2): 1–14.

Robinson, J. D., F. Vahedifard, and A. AghaKouchak. 2017. "Rainfall-triggered slope instabilities under a changing climate: Comparative study using historical and projected precipitation extremes." *Can. Geotech. J.* 54 (1): 117–127.

Royal Society. 2014. *Resilience to extreme weather.* Royal Society Science Policy Centre Rep. 02/14. London: Royal Society.

Scharlemann, J., E. Tanner, R. Hiederer, and V. Kapos. 2014. "Global soil carbon: Understanding and managing the largest terrestrial carbon pool." *Carbon Manage.* 5 (1): 81–91.

Shimozono, T., and S. Sato. 2016. "Coastal vulnerability analysis during tsunami-induced levee overflow and breaching by a high-resolution flood model." *Coastal Engineering* 107: 116–126.

Smith, S. V., W. H. Renwick, J. D. Bartley, and R. W. Buddemeier. 2002. "Distribution and significance of small, artificial water bodies across the United States landscape." *Sci. Total Environ.* 299 (1–3): 21–36.

Urbano, M., et al. 2013. *Building resilience: Integrating climate and disaster risk into development.* World Bank Group experience: Main report. Washington, DC: World Bank.

USACE (US Army Corps of Engineers). 2009. *Study findings and technical report. Alaska baseline erosion assessment.* Elmendorf, AK: USACE.

USACE. 2013. "The US Army Corps of Engineers: A brief history." Ebook. Accessed May 22, 2019. https://www.usace.army.mil/About/History/Brief-History-of-the-Corps/.

USACE. 2019a. "National inventory of dams." Accessed May 22, 2019. https://nid-test.sec.usace.army.mil/ords/f?p=105:1.

USACE. 2019b. "National levee safety website." Accessed May 22, 2019. https://www.usace.army.mil/National-Levee-Safety/.

USGCRP (US Global Change Research Program). 2018. Vol. 2 of *Impacts, risks, and adaptation in the United States: Fourth National Climate Assessment,* edited by D. R. Reidmiller, C. W. Avery, D. R. Easterling, K. E. Kunkel, K. L. M. Lewis, T. K. Maycock, et al. Washington, DC: USGCRP.

Vahedifard, F., A. AghaKouchak, E. Ragno, S. Shahrokhabadi, and I. Mallakpour. 2017a. "Lessons from the Oroville Dam." *Science* 355 (6330): 1139–1140.

Vahedifard, F., A. AghaKouchak, and J. D. Robinson. 2015. "Drought threatens California's levees." *Science* 349 (6250): 799.

Vahedifard, F., F. H. Jasim, F. T. Tracy, M. Abdollahi, A. Alborzi, and A. AghaKouchak. 2020. "Levee fragility behavior under projected future flooding in a warming climate." *J. Geotech. Geoenviron. Eng.* 146 (12): 04020139.

Vahedifard, F., J. D. Robinson, and A. AghaKouchak. 2016. "Can protracted drought undermine the structural integrity of California's earthen levees?" *J. Geotech. Geoenviron. Eng.* 142 (6): 02516001.

Vahedifard, F., S. Sehat, and J. V. Aanstoos. 2017b. "Effects of rainfall, geomorphological and geometrical variables on vulnerability of the lower Mississippi River levee system to slump slides." *Georisk: Assess. Manage. Risk Eng. Syst. Geohazards* 11 (3): 257–271.

Vardon, P. J. 2015. "Climatic influence on geotechnical infrastructure: A review." *Environ. Geotech.* 2 (3): 166–174.

Vicuña, S., M. Hanemann, and L. Dale. 2006. *Economic impacts of delta levee failure due to climate change: A scenario analysis.* Berkeley for the California Energy Commission, PIER Energy-Related Environmental Research. CEC-500-2006-004. Berkeley, CA: Univ. of California.

Volpi, E. 2019. "On return period and probability of failure in hydrology." *Wiley Interdiscip. Rev.: Water* 6 (3): e1340.

Wahl, T., S. Jain, J. Bender, S. D. Meyers, and M. E. Luther. 2015. "Increasing risk of compound flooding from storm surge and rainfall for major US cities." *Nat. Clim. Change* 5 (12): 1093–1097.

Wahl, T., P. J. Ward, H. C. Winsemius, A. AghaKouchak, J. Bender, I. D. Haigh, et al. 2018. "When environmental forces collide." *EOS* 99.

Wang, R.-Q., M. T. Stacey, L. M. M. Herdman, P. L. Barnard, and L. Erikson. 2018. "The Influence of Sea Level Rise on the Regional Interdependence of Coastal Infrastructure." *Earth's Future* 6 (5): 677–688.

Wang, Y., C. Chen, J. Wang, and R. Baldick. 2016. "Research on resilience of power systems under natural disasters—A review." *IEEE Trans. Power Syst.* 31 (2): 1604–1613.

Wehner, M. F., J. R. Arnold, T. Knutson, K. E. Kunkel, and A. N. LeGrande. 2017. "Droughts, floods, and wildfires." In Chap. 8 in Vol. 1 of *Climate science special report: Fourth National Climate Assessment*, edited by D. J. Wuebbles, D. W. Fahey, K. A. Hibbard, D. J. Dokken, B. C. Stewart, and T. K. Maycock, 231–256. Washington, DC: US Global Change Research Program.

World Bank. 2013. *Building resilience: Integrating climate and disaster risk into development.* Washington, DC: World Bank.

WHO (World Health Organization). 2017. *Strengthening resilience: A priority shared by health 2020 and the sustainable development goals*, edited by N. Satterley. Venice, Italy: WHO.

WSJ (Wall Street Journal). 2019. "Seaports add protection as ocean levels rise." *Wall Street Journal*, February 12, 2019, B-2.

Zhang, W., G. Villarini, G. A. Vecchi, and J. A. Smith. 2018. "Urbanization exacerbated the rainfall and flooding caused by hurricane Harvey in Houston." *Nature* 563 (7731): 384–388.

Zindlor, E. 2019. "2019 Factbook launch." BloombergNEF, L.P. Accessed February 13, 2019. https://www.bcse.org/factbook/.

CHAPTER 4

Considerations for Prioritization

Although traditional engineering design seeks to maintain asset strength and safety of life for a target *design event*, the longer-term fragility arising from climate change degradations has not yet been enforced by design standards. Not only is there a need to improve the capacity for emergency response and to speed up the recovery process, but also it is necessary to consider the major capital investments that could minimize or avoid disaster. Recent disasters have highlighted that the loss of energy, transportation, and telecommunications affects *normal life* in the short term and have both short- and long-term economic consequences. However, over a longer time frame, major disruptions, such as loss of water supply because of increased droughts, reduced groundwater, or reduced snowpack, will have irretrievable health consequences that can be avoided or mitigated by taking right actions now and by properly prioritizing these actions. In addition, the cascading effects of disruptive events from one sector, through the vast network of interdependencies described in Chapter 3, to multiple other sectors can be catastrophic, particularly as these interdependencies become more vital in the face of climate change. A combination of the consequences of a disruptive event on each sector, along with the cascading effects because of sector interdependencies, should be considered in any prioritization scheme.

Many prioritization schemes are based on an analysis of the relative merits and drawbacks of improvements or the tangible impacts of failure. These could be measured as a proportion of the population affected, direct costs, indirect costs (e.g., lost working hours), health impacts, vehicle hours wasted, and so on. However, such information is often proprietary and not publicly available or is not comparable across different infrastructure categories or different locations. As a result, other factors must be taken into account, including the current status of each sector's adaptation efforts, the timeline by which such efforts could be planned and implemented, the significance of each sector as it pertains to a particular region (e.g., navigation is not important in all watersheds), and the perspective of the analyst (public, private, government), among others. The vulnerability of different sectors also varies regionally. For instance, transportation in coastal regions has a different degree of exposure than transportation inland, whereas available funds for maintenance can also affect the regional sensitivity and vulnerability. Similarly, the level of redundancy in many infrastructure systems differs between rural and urban environments as well as across the United States. Thus, in one location, a combination of exposure, a lack of redundancy, and sensitivity to disruptive events may lead to energy infrastructure being the priority for investment and adaptation action, whereas in another location, future scarcity may elevate drinking water infrastructure as the main priority.

As organizations grapple with the importance and scale of undertaking actions of adaptation and resilience, several have derived methods to assess where to begin. The preceding chapters set the context by which prioritization of sectors can be informed based on funding for, and implementation of, climate change adaptation efforts. The contents were gleaned from various types of literature and from each author's particular expertise and experience and can be used as one source of information. This chapter is presented as a guide to potential approaches and questions that have been successfully used both domestically and internationally. The alternatives presented here do not comprise a comprehensive list of all possible prioritization schemes and questions but only provide a starting point from which to frame the discussion. We first discuss some of the other constraints faced by decision-makers, followed by a discussion of the possible ways to frame a prioritization scheme along with some examples.

4.1 DECISION-MAKING CONSTRAINTS

4.1.1 Policy Engagement

Policy interactions with decision-makers occur at multiple levels, from corporate management to international or national government. There is a risk that without clear guidance from governmental policy-makers, adaptation assessments and vulnerability analyses will remain as reports that sit on the shelf and are not implemented. Constructing and maintaining resilient infrastructure will require considerable changes to existing financial and regulatory policies (Lempert et al. 2018). Although design standards are commonly set at a higher level by national and international standard setting organizations, individual state and local governments have the ability to increase these minimum requirements (ASCE CACC 2018, CSIWG 2018). A concentrated effort to update design codes is required to meet the challenges already listed in this report. However, the design standard revision cycle can be protracted, putting the onus on engineers, building owners, and agencies to advocate for enhanced designs (ASCE CACC 2018). Given the increased costs associated with resilient construction, designers and contractors are reluctant to propose enhancements that go beyond the minimum requested by the client. However, there is a trickle-down effect to the private sector of improved minimum standards in design as federal agencies and some state and local municipalities take the lead (ADOT 2020, FHWA 2015, NIAC 2015).

The interconnectedness of infrastructure and the potential for greater cascading impacts is highly dependent on the interactions of different governing and operating agencies, and past policies and decisions. Past policies may also have led to *infrastructure lock-in* (Markolf et al. 2018), not only increasing the negative consequences but also hindering future adaptive actions. Although the probability of landfalling hurricanes is higher with a changing climate, the influence of land use and building practices, environmental degradation, and fiscal incentives all contribute substantially to the potential consequences (Doss-Gollin et al. 2020). Integrating the management and decision-making responsibility across communities, agencies, and industries can lead to enhanced resilience and improved social justice (Thaler et al. 2018). Conversely, any efforts to improve resilience in one sector may be severely hampered by failing to recognize the influence of other systems or business interdependencies. Thus, policies and practices also need to address the management of interdependent systems, incorporate a multisector perspective, and acknowledge the societal contexts (Clarke et al. 2018). A multisector approach can also broaden the scope for benefits across infrastructure projects. For instance, introducing rain gardens and green roofs into the urban plan will help to provide mental health benefits, mitigate pluvial flooding and the urban heat island effect, and enhance transportation safety.

As discussed later, the insurance and finance industries not only offer a potential mechanism for funding adaptation actions but can also encourage a higher design standard by increasing insurance premiums for noncompliant infrastructure (Simmons et al. 2018). Not only is there greater interest from financial institutions to protect their assets, but also aligning the interests of investors with those of operators, insurers, and regulators will deliver greater resilience (Florin and Sachs 2019). Greater interdisciplinary knowledge and data sharing leveraging the improvements in infrastructure digitalization can assist operators in developing redundancy and managing their assets (ICE 2020).

4.1.2 Coherence with Climate Change Adaptation Plans

The first step in demonstrating concern for the potential impacts of climate change and mapping out adaptation strategies to combat these future impacts is the development of a federal-, state-, and/or city-level climate change adaptation plan. Such plans represent official statements or strategies regarding government policy on this issue and should also provide decisive and clear descriptions of the objectives to be achieved to counter climate change. The information included in a climate change adaptation plan should include the specific action items that will be implemented to reduce overall vulnerability, how these action items will be implemented, and the time frame over which each action will occur. Metrics that measure the level of implementation of the suggested action items within a plan, as well as the level of success that such action items have had on lowering overall vulnerability to climate change, should be available. Prioritization of specific projects should relate to how well they fit within the overall objectives of the plan and the extent to which they will contribute to improving the metrics used to evaluate the implementation of the plan.

An example of a qualitative prioritization scheme that considers the objectives previously discussed is shown in Figure 4-1. This scheme was developed within the United Kingdom by the Adaptation Sub-Committee under the Committee on Climate Change (2017) to assess climate change risk within the United Kingdom after adoption of the National Adaptation Programme. The assessment ranks adaptation priorities for projects based on sector according to whether that sector is considered within an adaptation plan, if a plan exists at all, and whether progress has been made in addressing vulnerability to climate change within the proposed project's sector. Each colored box represents different combinations of levels of consideration within a plan and levels of progress made, with red representing minimal or no consideration and little if any progress being made; lists of sectors/priorities within each combination of consideration and progress are listed within each box.

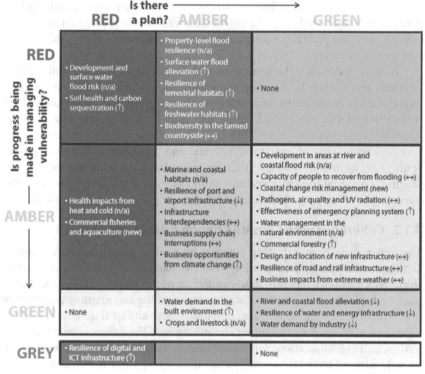

Figure 4-1. Qualitative assessment of the consideration of various sectors within a climate change adaptation plan and their respective level of progress made; low levels of each are represented by red and brown, whereas higher levels are represented by yellow and green. Sectors falling within each level are listed.

Source: Committee on Climate Change (2017).

4.1.3 Sustainable Development and Social Inequality

The stakeholder groups that are most often forgotten in the planning process are the communities that will be served by the infrastructure. Without full community engagement, there is a risk of perpetuating social and economic inequities through a lack of engagement of those most likely to be disadvantaged by disruptive events (Thaler et al. 2018). Differences in culture and levels of trust in government-appointed decision-makers have a considerable impact on communities' acceptance of infrastructure improvements. Understanding the differences in how people plan to, and actually, respond to hazards is critical to ensure that infrastructure planning and design is resilient (Attems et al. 2020, Morss et al. 2011). Further, those communities will have the underlying knowledge of the locality and the likelihood of particular adaptation strategies performing well (Bonham-Carter et al. 2014).

With this in mind, an often neglected factor when prioritizing infrastructure investments is consideration of the extent to which a project promotes sustainable development and reduces social inequality. The Movement Strategy Center argues that resilience to climate change requires a holistic effort that encompasses the eradication of inequality and the unsustainable use of resources (Pathways to Resilience Partners 2015). It is often the case that the most disadvantaged in society have the greatest exposure to disruptive events and climate change impacts. Addressing this reality requires a commitment toward investing in infrastructure projects that facilitate improvements in safety, health, and economic opportunities for the disadvantaged. CSIWG (2018) makes the point that any no- or low-regret policy in terms of climate change adaptation and future investment must be aware of the infrastructure investment gaps that increase the vulnerability of infrastructure to future climate conditions in which it is meant to serve. As is the case in most locations, infrastructure in most need of investment tends to be unevenly distributed socioeconomically, with disadvantaged communities (e.g., low income and minorities) often having to deal with such issues as dilapidated schools and other types of buildings, the least convenient access to reliable mass transportation and communication networks, poorly maintained drinking water and wastewater conveyance systems, and so on. One of the main contributors to this lack of investment is the fact that said communities are typically not involved or invited to engage in the planning and decision-making stages of projects that should address these issues. Because of the fact that the added vulnerability from climate change is also disproportionately biased toward these communities as well and that these communities have the most limited resources available to deal with climate change, the inequality in the infrastructure investment gap will be further exacerbated in the future. Engagement of communities affected, which includes eliminating the persistent institutional racism that currently exists, is vital to being able to address both the current infrastructure investment gaps and future investment gaps that will be created because of climate change impacts. Creating jobs and providing more opportunities for workforce training should also be part of any solution or project that is meant to address social inequality issues.

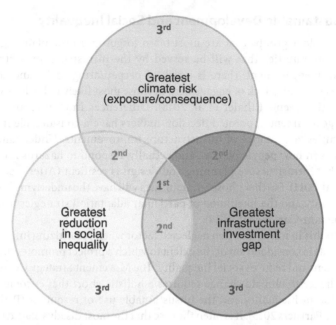

Figure 4-2. Infrastructure investment scheme based on climate risk, potential reduction in social inequality, and addressing a current infrastructure investment gap.
Source: Fig. 4.5 of CSIWG (2018).

With the aforementioned issues and solutions in mind, the climate-safe path prioritizes infrastructure investment using the scheme illustrated in Figure 4-2 (CSIWG 2018). The lowest priority is given to those projects that address only one of the major questions regarding risk to the impacts of climate change, a reduction in social inequality, or addressing an infrastructure investment gap; this priority level is illustrated in Figure 4-2 by areas labeled as 3rd. The next priority level includes projects that will alleviate any two of these issues and is illustrated by the overlap between any two circles (labeled as 2nd) in Figure 4-2. Projects deemed as the highest priority are those that address all three issues and are illustrated by the overlap of all circles in the middle of Figure 4-2 (labeled as 1st).

4.1.4 Economics

With unlimited budgets, the case for adapting to future climate is straightforward. However, such a luxury is rare, as maintenance and operation costs often consume a vast proportion of any annual infrastructure budget. Thus, decision-makers often have to juggle the need to plan for and adapt to climate change against maintaining functionality at the current standards (NASEM 2020). Benefit–cost analysis (BCA) tools are frequently used to select the most cost-effective and beneficial infrastructure (Multihazard Mitigation Council 2018), but decision-makers lack the financial information needed to assess the relative merits of incorporating adaptation for climate change (Lempert et al. 2018). Past infrastructure failures can act as a guide for the avoided costs engendered by resilient infrastructure construction (Hay et al. 2016), but greater clarity and data are also needed, particularly on the efficacy of adaptation mechanisms (McEvoy et al. 2010).

The economic justification for incorporating adaptation measures cannot be realized simply by imposing future climate conditions on a specific asset. Chapter 1 discussed *nuisance flooding* and the difficulty in assessing the economic implications for road users, for instance. Not only are there direct maintenance costs associated with repairing flood damage and indirect costs associated with delays or increased travel (Pregnolato et al. 2017), but also there is the additional maintenance burden from adverse loading on alternate routes. The anticipated design life can have a considerable impact on the eventual decision, where the risks posed by interannual variability outweigh those posed by the climate (Doss-Gollin et al. 2019). In short, BCA should incorporate the full life cycle and operational activities of the proposed asset and consider the wider network and the potential to cause or respond to cascading failure (NASEM 2020).

At a broader scale, attention has been paid to quantifying the likely return on investment for resilient and sustainable infrastructure. For instance, by 2050, improved infrastructure that is aligned with inclusive urban development could result in $17 trillion economic savings (Mountford et al. 2018). If one estimates the exposure costs for coastal communities simply from sea-level rise, hurricanes, and other storms, assuming that other factors such as the urban density remain constant, the annual average costs could increase by $2 to $3.7 billion in the coming decade and a further $6 to $12 billion in the 2040s (Hsiang et al. 2014). However, there are few examples of the economic impacts at the small community level, with analyses mostly undertaken by large cities, counties, or states (Arcadis 2017, Bonham-Carter et al. 2014, GHD 2014, Hayhoe et al. 2008).

Although the benefit of adapting infrastructure for climate change will lead, eventually, to society saving between $4 and $11 for each $1 spent (Multihazard Mitigation Council 2018), investment is required up front. With an investment shortfall of $2 trillion over the next 10 years (ASCE 2016) and the financial constraints already faced by the local and municipal authorities who will likely undertake these actions, targeted involvement of the private sector is necessary (Mountford et al. 2018). Private investment already accounts for around half of global infrastructure investment, and it is estimated that this could increase considerably with the benefit of sound policies and clear regulations (Bielenberg et al. 2016). Thus, the funding mechanisms already exist to achieve resilience but will need iteration and updating to meet the needs of different sectors (ICE 2020). It is also important to consider the role that can be played by insurance. Insurance can cover the residual risks that cannot be economically engineered away (Nikellis and Sett 2020, ORR 2018) and also create positive feedback by requiring higher design standards to reduce the risks transferred to insurance agencies (Florin and Sachs 2019).

Methods for estimating the relative merits of adaptation measures have varying degrees of complexity and are sometimes conflated with the costs of climate change mitigation, where, say, the resilience of energy infrastructure also brings avoided carbon emissions. Economic analyses of the impacts of climate change were first quantified in the early 1990s using cost–benefit impact assessment models (IAMs) (Nordhaus 1994). However, as the focus of IAMs is primarily on emission reduction (mitigation) rather than adaptation, more recent developments have combined climate projections, econometric research, and private sector risk models to assist with local decision-making (e.g., the spatial empirical global to local assessment system; Hsiang et al. 2014). Although such models are computationally expensive and rely on proprietary information, some of the indices, such as changes in the number of hours worked or electricity demand, can be used to estimate the costs avoided by different adaptation measures. Such estimates are used in cost-effectiveness analyses (Hunt 2017).

FEMA's BCA toolkit has recently been updated to incorporate multiple weather- and climate-related hazards in addition to infrastructure failure or dam breaks (FEMA 2019), with several options for cost estimation. Other BCA tools for major infrastructure works can be modified to accommodate the anticipated changes in risk or exposure related to adaptation actions such as those developed by the Federal Highways Authority (Filosa et al. 2018), the National Cooperative Highways Research Program (NASEM 2020), or the US Army Corps of Engineers (Brekke et al. 2009). Some research is also underway in the water sector to quantify the business impacts of climate change (Wasley et al. 2020).

Detailed assessments may not always be appropriate. For instance, where the anticipated life of the structure is short, the changes anticipated from climate change may be dwarfed by other changes such as land use or population growth. Thus, three different levels of analysis have been suggested to encompass different complexities of projects (NASEM 2020)

- Low regrets sketch analysis: A simple analysis of whether the adaptation would be beneficial for the public or whether the current situation is viable for the future. This requires very little data and depends on vast simplification.

- Climate resilience analysis: A broad-brush assessment of the range of adaptation measures that can increase resilience through reducing ongoing or projected losses. This requires moderate levels of information, and may be an effective way of incorporating incremental design adaptations and monitoring climate effects.

- Detailed engineering resilience analysis: This is the most appropriate for site-or project-specific assessments and entails considerable data collection. It is most relevant for long life projects such as new highways or water storage reservoirs.

4.2 BALANCING FINANCIAL, ECONOMIC, SOCIAL, AND ENVIRONMENTAL COSTS AND BENEFITS

The financial returns of investment in climate change adaptation are often the primary focus of the private perspective. In contrast, the social, economic, and environmental benefits of adaptation focus on improving health, protecting the environment, encouraging diffuse economic activity, and avoiding loss of life. However, the benefits derived by a project in relation to one of the financial, economic, social, or environmental gains are often offset by costs within the other areas. Thus, it can be difficult to weigh the resulting costs against the expected benefits when determining the overall impact of a project. The Center for Climate and Energy Solutions (C2ES) stresses the need not only to prioritize adaptation projects according to their ability to withstand changing conditions but also to use green infrastructure such as renewable energy and energy and stormwater conservation measures (C2ES 2018). The installation and use of green infrastructure are not always straightforward in balancing overall benefits versus costs. Solar farms, for example, represent one potential solution that promotes the use of renewable energy for a community while reducing greenhouse gas emissions. Yet, they also pose a potentially negative environmental impact arising from the loss of rural land required to sustain such a project. To determine the net benefit and relative priority of such a project, a prioritization scheme that takes all this into account must be devised. Examples of such schemes have been described by the World Bank [World Bank Infrastructure Prioritization Framework (WBIPF)] (Marcelo et al. 2016) and the United Nations Development Program (Kotchen 2011).

The WBIPF, for instance, prioritizes infrastructure using a multicriteria approach accounting for project-level social, environmental, financial, and economic factors. In addition to the traditional budgetary concerns, the approach considers national policy objectives, aspirations for long-term development, social inequality, and environmental sustainability. All these factors are summarized into two indices: a Social and Environmental Index (SEI) and a Financial and Economic Index (FEI). The SEI specifically accounts for project benefits related to improved access to public services and increased job opportunities during project construction and execution while also recognizing the environmental costs related to removing vegetation, polluting natural environments, and relocating families or communities. The FEI accounts for not only financial profitability and economic value of the project but also indirect multiplier or interdependency effects on other sectors.

An example using the SEI and FEI to assess project prioritization is shown in Figure 4-3 for Vietnam Transport Projects (Marcelo et al. 2016). In this example, the SEI was computed using five indicators related to jobs, the number of beneficiaries, land-use repurposing, environmental and cultural risks, and greenhouse gas emissions. The FEI used indicators to account for the financial internal rate of return, multiplier effects, and a project's implementation risk to estimate the project's location with regard to areas designated as priority economic zones and their relative alignment with existing infrastructure in the area. The resulting normalized scores identify higher priority projects as those with FEI and SEI scores nearer to 100. Figure 4-3 illustrates the ranking of individual Vietnam Transport projects.

The dashed lines in Figure 4-3 represent the budget constraints for a particular sector. The horizontal dashed line represents the limit of those projects that can be financed within the budget based on the SEI score alone, whereas the vertical

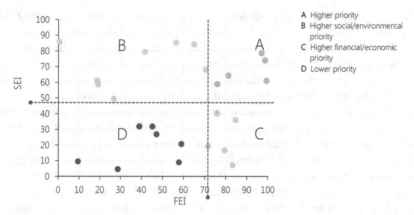

Figure 4-3. Example using the SEI and FEI to assess prioritization of Vietnam Transport Projects. The dashed lines represent budget restraints in relation to the SEI (horizontal) and the FEI (vertical). Circle colors represent levels of priority (green = highest; red = lowest).

Source: Fig. 5 in Marcelo et al. (2016). Permission to reproduce under the Creative Commons License: https://creativecommons.org/licenses/by/3.0/igo/legalcode.

dashed line represents the limit of those projects that can be financed within the budget based on the FEI score alone. The top-right quadrant that forms as a result of adding the two budget-constraint lines contains the highest priority projects (represented by green dots) that fall within the budget constraints when considering both the SEI and FEI indices. Projects falling within the budget constraints when considering only one of the indices (yellow dots) are considered lower priority overall even though they are a high priority for at least one of the indices. Projects located in the lower-left quadrant (red dots) are considered the lowest priority because of the fact that they have low SEI and FEI scores and do not fall within budget constraints.

Kotchen (2011) performs another type of quantitative analysis by comparing the costs and social benefits of implementing climate-proofing infrastructure. For a specific project, the incremental increase in costs associated with climate proofing (marginal cost or MC) and associated social benefits (marginal social benefit or MSB) are estimated. The area under each curve in Figure 4-4 represents the total cost [TC in Figure 4-4(a)] and total social benefit [TSB in Figure 4-4(b)].

The goal is to determine the optimum level of climate proofing by comparing MC and MSB to determine the point at which an increase in MCs exceeds the increase in marginal benefits (MSBs), shown as point Q* in Figure 4-5(a). Additional climate proofing above this level results in marginal costs that exceed the associated marginal benefits and is, thus, not economically efficient. The net social benefit (NSB) is determined from the area bounded by the MSB and MC curves and point Q* [blue shaded area in Figure 4-5(a)]. Alternatively, marginal net societal benefits (MNSBs) are calculated by subtracting MC from the MSB curve, forming the curve in Figure 4-5(b). An economically efficient level of climate

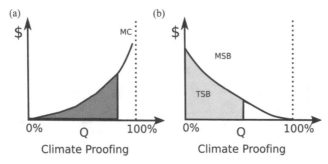

Figure 4-4. Plots of the (a) marginal cost, (b) marginal benefit curves. TC and TSB represent the total cost and total social benefit, respectively, associated with a certain level (Q) of climate proofing.

Source: Figs. 9.2 and 9.3 of Kotchen (2011).

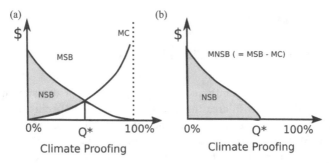

Figure 4-5. Plots of (a) MSB, (b) MNSB. The shaded blue areas labeled NSB in each plot represent the net social benefit.

Source: Fig. 9.5 of Kotchen (2011).

proofing, therefore, produces a positive value of the MNSB, with the optimum level of climate proofing achieved when the MNSB reaches 0. The resulting NSB is represented by the area under the curve in Figure 4-5(b). Additional climate proofing above this level will result in a negative MNSB and reduce the NSB.

The final step is to prioritize climate-proofing allocations for multiple projects when budget constraints do not allow for the optimum level of climate proofing in each project. The objective is to allocate resources that achieve maximum total net social benefits across all projects for a given total budget. An example in Figure 4-6 illustrates the process of maximizing the NSB across electrification (e.g., bolstering power lines) and transportation (e.g., flood proofing) projects. By equating the MNSB between the two sectors, Q'_A and Q'_B provide the most efficient levels of climate proofing for each project for the total budget, resulting in revised levels of climate proofing and the NSB for each project. The advantage of this method is that it helps to prioritize projects in terms of the social net benefits, even when these cannot be fully quantified, and suggests the relative levels of climate proofing to achieve the maximum NSB.

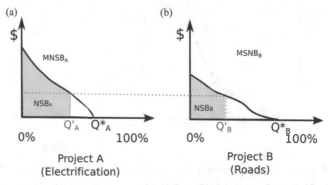

Figure 4-6. Example of equating MNSB between projects in two different sectors (energy and transportation). Q'_A and Q'_B represent the levels of climate proofing in each sector, respectively, that maximize total NSB (shaded area) under budget constraints (dotted line).

Source: Fig. 9.6 of Kotchen (2011).

4.3 SYNTHESIS OF PRIORITIZATION STEPS

Drawing from the aforementioned examples, a series of decision steps have been developed to assist in creating a sector- or location-specific prioritization framework. The decision steps are informed by the United Nations Environment Programme Multi-Criteria Assessment for climate change prioritization tool and include the following (adapted from Republic of Ghana Ministry of Finance 2016)

1. Understand the decision context;
2. Identify options for adaptation;
3. Identify the list of criteria and indicators;
4. Scoring, weighting, and interpreting results; and
5. Conduct a sensitivity analysis.

   Although not every prioritization scheme will specifically include each step, they will include most of them and will follow this framework to a large extent.

4.3.1 Step 1: Understand the Decision Context

The first step places the proposed adaptation project(s) within the context of national and state climate policy and action plans. Thus framed, the prioritization will include specific development and climate change priorities and strategies along with a review of current and future projects within the sector of the project(s) under consideration.

4.3.2 Step 2: Identify Options for Adaptation

The next step is to create a comprehensive list of technologies that address the specific objectives identified within the sector(s) or the project(s) under consideration. This step requires a thorough literature review prior to committing to one or more specific technologies or design solutions. The literature, here, includes national or local planning documents, assessments of climate change vulnerability, and/or risk within the current region of influence and nearby areas. This step should also include engagement with relevant experts and affected stakeholders, including those from disadvantaged groups, to brainstorm other pertinent adaptation options. After creating a list of potential technologies, the costs and benefits, operation and maintenance considerations, application potential and efficiency, and potential barriers to full implementation need to be considered.

4.3.3 Step 3: Identify the List of Criteria and Indicators

The next step defines the questions or criteria that will be used to measure the potential performance of each adaptation option. Criteria can measure the impact of technology in terms of increased resilience and decreased vulnerability to future climate change impacts, coherence with national and local climate change adaptation and development plans, promotion of sustainable development through reductions in social inequality, protection of the environment, operational and maintenance costs, and institutional barriers. Table 4-1 provides a list of criteria that have been considered in prior prioritization schemes, including those related to impacts on climate change adaptation (ASCE CACC 2018, C2ES 2018, CSIWG 2018, Community Resilience Program NIST 2017, Espinet and Rozenberg 2018, Republic of Ghana Ministry of Finance 2016, The Cities of Eugene and Springfield Oregon 2014; DOT 2015), sustainable social and environmental development and costs (C2ES 2018, Community Resilience Program NIST 2017, CSIWG 2018, Espinet and Rozenberg 2018, Kotchen 2011, Marcelo et al. 2016, Multihazard Mitigation Council 2018, Republic of Ghana Ministry of Finance 2016), and financial and economic cost–benefit analysis (Community Resilience Program NIST 2017, Espinet and Rozenberg 2018, Marcelo et al. 2016, Multihazard Mitigation Council 2018, Republic of Ghana Ministry of Finance 2016).

4.3.4 Step 4: Scoring, Weighting, and Interpreting the Results

After determining the criteria that are important within a particular country, region, or locality, the next step is to devise a scoring method by which to prioritize the schemes. Such a method could be qualitative or quantitative and should be illustrated in a simple format that allows effective communication of the ranked priorities. The choices of qualitative- or quantitative-based prioritization schemes, questions to be, and final presentation of the results depend on the specific country, region, or locality, and sectors of interest for the proposed adaptation activities. Regardless of the scheme selected and questions addressed, maximum transparency of the prioritization scheme and communication of the results is paramount.

A quantitative method begins with estimating the relative importance of each criterion via measurable variables or indicators. The use of indicators is contingent on data availability and resolution. Multiple indicators may be used

Table 4-1. Examples of Questions Addressed and Prior Studies That Have Incorporated These Questions into Their Respective Prioritization Schemes.

Question	Study
Is there a climate change adaptation plan?	• Committee on Climate Change (2017)
Are there interdependencies and cascading/ external impacts/ benefits?	• Committee on Climate Change (2017) • Community Resilience Program NIST (2017) • UNDP (Kotchen 2011)
What are the greatest sensitivities, exposures, consequences, and potential risk reduction to climate change indicators/impacts?	• ASCE Method of Practice 140 Chapter 7 (ASCE CACC 2018) • C2ES (2018) • Climate Change Project Prioritization Tool (Republic of Ghana Ministry of Finance 2016) • CSIWG (2018) • Espinet and Rozenberg (2018) • Vulnerability Assessment Scoring Tool (VAST) (DOT 2015)
What are the social benefits and costs?	• C2ES (2018) • Climate Change Project Prioritization Tool (Republic of Ghana Ministry of Finance 2016) • Community Resilience Program NIST (2017) • CSIWG (2018) • Espinet and Rozenberg (2018) • Infrastructure Prioritization Framework (Marcelo et al. 2016) • UNDP (Kotchen 2011)
Where is the greatest infrastructure investment gap?	• Community Resilience Program NIST (2017) • CSIWG (2018)
What are the environmental benefits and costs?	• Climate Change Project Prioritization Tool (Republic of Ghana Ministry of Finance 2016) • Community Resilience Program NIST (2017) • Infrastructure Prioritization Framework (Marcelo et al. 2016) • UNDP (Kotchen 2011)
What are the financial and economic costs and benefits?	• Climate Change Project Prioritization Tool (Republic of Ghana Ministry of Finance 2016) • Community Resilience Program NIST (2017) • Espinet and Rozenberg (2018) • Infrastructure Prioritization Framework (Marcelo et al. 2016) • McDaniels et al. (2015)

(Continued)

Table 4-1. *Examples of Questions Addressed and Prior Studies That Have Incorporated These Questions into Their Respective Prioritization Schemes. (Continued)*

Question	Study
Occupancy considerations	• ASCE Method of Practice 140 Chapter 7 (ASCE CACC 2018)
Service life	• ASCE Method of Practice 140 Chapter 7 (ASCE CACC 2018)
Ease of implementation (e.g., political/ institutional barriers, time required, number of agencies involved)	• Climate Change Project Prioritization Tool (Republic of Ghana Ministry of Finance 2016) • Community Resilience Program NIST (2017) • McDaniels et al. (2015)
Incorporation of green infrastructure	• C2ES (2018)

to measure a criterion, in which case each indicator also needs to be weighed. Finally, the overall scoring system joins each weighed indicator with a composite score for the relevant criterion and combines the scores for each criterion with one score of priority. Examples demonstrating the implementation of this methodology include the US Army Corps of Engineers (USACE 2019) Watershed Vulnerability Assessment Tool, which uses indicators related to projected hydrological, ecological, social, and other types of factors to prioritize USACE projects throughout the United States; the Global Facility for Disaster Reduction and Recovery guide book for practitioners, which provides a specific example of adaptation project prioritization for the road network in Belize (Kappes et al. 2017); the Republic of Ghana Climate Change Project Prioritization Tool (Republic of Ghana Ministry of Finance 2016); and the US Department of Transportation (DOT 2015) Vulnerability Assessment Scoring Tool (VAST).

4.3.5 Step 5: Sensitivity Analysis

The final step of the prioritization process considers the many remaining uncertainties, especially if a quantitative analysis has been performed. Uncertainties can be estimated by assessing the dependence of the sensitivity of project rank to factors that are characterized by measurement errors, subjective judgment, or limited knowledge. For example, the weights assigned to each criterion indicator are subjective and rely heavily on the specific region under consideration and the needs and objectives of competing stakeholders. For instance, a coastal region would likely place much more importance or higher weights on indicators related to navigation and sea-level rise than a region or municipality that is located further inland. The results of the sensitivity analysis should then be applied to the tentative project ranks to produce the final rankings.

4.4 GENERALIZED PRIORITIZATION OF US INFRASTRUCTURE

The assessments of likely future changes in disruptive events and their manifestation as impacts on individual infrastructure sectors are used to illustrate a potential prioritization ranking. In the absence of substantial data on the costs, benefits, or distribution of assets, a qualitative assessment is used based on the authors' expertise and judgment. The prioritization is considered at a national level, ignoring the differences in regional changes outlined in Chapter 2 and the different levels of infrastructure resilience or fragility in each state. The questions considered are those used in other high-level assessments (ASCE CACC 2015, C2ES 2018, Committee on Climate Change 2017, Community Resilience Program NIST 2017, Republic of Ghana Ministry of Finance 2016, DOT 2015): Are there interdependencies and cascading/external impacts/benefits? What are the sensitivities, exposures, or consequences flowing out from climate change impacts?

Figure 4-7 illustrates the qualitative assessment of the consequences for each sector from a disruptive event (i.e., the level of sensitivity plus exposure) and the level of interdependency among sectors and the potential to cause cascading failure. In this instance, impacts are considered from the perspective of the short-term response to an increased frequency or magnitude of disruptive events. An alternate approach might consider the longer-term consequences of climate change that would benefit from actions now to avoid catastrophe later, such as exhaustion of water supplies in snowpack-driven regions. Each sector is represented by an icon, the position of which is determined by the potential impacts experienced by a sector because of a disruptive event (x-axis) and the interdependency or potential to cause cascading failure (y-axis). Priority increases from yellow to red and is given to those sectors exhibiting the highest levels of interdependency and potential short-term impacts. Given an overall infrastructure score of D+, indicating that infrastructure is of a poor standard and is showing deterioration (ASCE 2017), all sectors are vulnerable to disruptive events and prone to failure. Sector icons located in the yellow and orange portions of Figure 4-7 still require considerable investment but will have fewer cascading impacts relative to those in the red portion.

The analyses presented in Chapter 3 indicate that the interdependencies among the energy, roads, and bridges and flood protection sectors will result in widespread disruptions and prolonged economic, social, and environmental consequences. In contrast, disruptions to the navigation sector will have fewer cascading impacts as dependency on this sector is lower. Flood infrastructure, by definition, together with roads, bridges, energy, and water infrastructure, is often located at places with high exposure to disruptive events. The consequences of failure in any one of these sectors will have vast repercussions on the safety and welfare of society, for instance, through loss of access to vital medical support.

The results presented here are the considered opinion of the authors based on the available data and public information. Another combination of scientists, engineers, policymakers, and economists would likely arrive at a slightly different

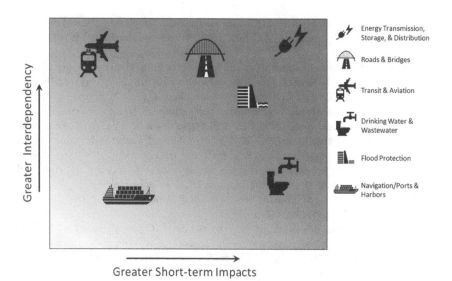

Figure 4-7. Qualitative prioritization of a US infrastructure sector's need for climate change adaptation based on the relative level of interdependency and the short-term impacts because of failure.

prioritization. Further, this example assessment is based on a national picture of infrastructure exposure and sensitivity and will vary considerably for more localized assessments. For instance, southeast US infrastructure is at the greatest risk from rising sea levels and increased flooding (USGCRP 2018), affecting primarily the roads and bridges sector. Similarly, all transportation infrastructure in the Midwest is highly vulnerable to increases in pluvial flooding (USGCRP 2018). In contrast, water supply is higher priority in California as a result of the high dependence on hydroelectric power, agriculture, and the poor state of physical assets (CSIWG 2018).

References

ADOT (Arizona Department of Transportation). 2020. *Asset management, extreme weather and proxy indicators pilot project.* Phoenix: ADOT.

Arcadis. 2017. *Resilient bridgeport. Benefit–cost analysis methodology report.* Bridgeport, CN: Arcadis.

ASCE. 2016. *Failure to act: Closing the infrastructure gap for America's economic future.* Washington, DC: ASCE.

ASCE. 2017. *Infrastructure report card: A comprehensive assessment of America's infrastructure.* Reston, VA: ASCE.

ASCE CACC (Committee on Adaptation to a Changing Climate). 2015. *Adapting infrastructure and civil engineering practice to a changing climate,* edited by J. R. Olsen. Reston, VA: ASCE.

ASCE CACC. 2018. *Climate-resilient infrastructure,* edited by B. M. Ayyub. Reston, VA: ASCE.

Attems, M.-S., T. Thaler, E. Genovese, and S. Fuchs. 2020. "Implementation of property-level flood risk adaptation (PLFRA) measures: Choices and decisions." *WIREs Water* 7 (1): e1404.

Bielenberg, A., M. Kerlin, J. Oppenheim, and M. Roberts. 2016. *Financing change: How to mobilize private sector financing for sustainable infrastructure.* New York: McKinsey.

Bonham-Carter, C., K. May, C. Thomas, A. Sudhalkar, B. Fish, A. Boone, et al. 2014. "Climate change and extreme weather adaptation options for transportation assets in the Bay Area Pilot project." Accessed July 15, 2020. https://www.fhwa.dot.gov/environment/ sustainability/resilience/pilots/2013-2015_pilots/mtc/final_report/index.cfm.

Brekke, L. D., J. E. Kiang, J. R. Olsen, R. S. Pulwarty, D. A. Raff, D. P. Turnipseed, et al. 2009. *Climate change and water resources management—A federal perspective.* U.S. Geological Survey Circular 1331. Washington, DC: US Geological Survey.

C2ES (Center for Climate and Energy Solutions). 2018. *Policy options for climate-resilient infrastructure.* Arlington, VA: C2ES.

Clarke, L., L. G. Nichols, R. Vallario, M. Hejazi, J. Horing, A. C. Janetos, et al. 2018. Vol. 2 of *Chapter 17: Sectoral interdependencies, multiple stressors, and complex systems. Impacts, risks, and adaptation in the United States: The Fourth National Climate Assessment.* Washington, DC: US Global Change Research Program.

Committee on Climate Change. 2017. *Progress in preparing for climate change.* Report to Parliament. London, UK: Committee on Climate Change.

Community Resilience Program NIST (National Institute of Standards and Technology). 2017. *Identifying and prioritizing closure of resilience gaps.* Guide Brief 13—Resilience Gaps—NIST Special Publication 1190GB-13. Gaithersburg, MD: NIST.

CSIWG (Climate-Safe Infrastructure Working Group). 2018. *Paying it forward: The path toward climate-safe infrastructure in California.* Report of the Climate-Safe Infrastructure Working Group to the California State Legislature and the Strategic Growth Council. Sacramento, CA: CSIWG.

Doss-Gollin, J., D. J. Farnham, M. Ho, and U. Lall. 2020. "Adaptation over fatalism: Leveraging high-impact climate disasters to boost societal resilience." *J. Water Resour. Plann. Manage.* 146 (4): 01820001.

Doss-Gollin, J., D. J. Farnham, S. Steinschneider, and U. Lall. 2019. "Robust adaptation to multiscale climate variability." *Earth's Future* 7 (7): 734–747.

DOT (US Dept. of Transportation). 2015. *Vulnerability assessment scoring tool.* Washington, DC: DOT.

Espinet, X., and J. Rozenberg. 2018. "Prioritization of climate change adaptation interventions in a road network combining spatial socio-economic data, network criticality analysis, and flood risk assessments." *Transp. Res. Rec.* 2672 (2): 44–53.

FEMA. 2019. "FEMA benefit–cost analysis (BCA) toolkit version 6.0 user guide." Accessed July 15, 2020. https://www.fema.gov/media-library-data/1571164308638-adf025324225 d699f7d9ee53bc618fa8/Version_6.0_User_Guide.pdf.

FHWA (Federal Highway Administration). 2015. *Transportation system preparedness and resilience to climate change and extreme weather events.* Order 5220. Washington, DC: FHWA.

Filosa, G., A. Plovnik, L. Stahl, R. Miller, and D. Pickering. 2018. *Vulnerability assessment and adaptation framework.* 3rd ed. Washington, DC: FHWA.

Florin, M.-V., and R. Sachs. 2019. "The role of insurance in critical infrastructure resilience." Accessed March 20, 2019. https://irgc.epfl.ch/wp-content/uploads/2019/09/ THE-ROLE-OF-INSURANCE-IN-CRITICAL-INFRASTRUCTURE-RESILIENCE-Article.pdf.

GHD (Gutteridge, Haskins & Davey Limited). 2014. *District 1 climate change vulnerability assessment and pilot studies.* FHWA Climate Resilience Pilot Final Report. Washington, DC: FHWA.

Hay, J. E., D. Easterling, K. L. Ebi, A. Kitoh, and M. Parry. 2016. "Conclusion to the special issue: Observed and projected changes in weather and climate extremes." *Weather Clim. Extreme.* 11: 103–105.

Hayhoe, K., D. Wuebbles, J. Hellmann, B. Lesht, K. Nadelhofer, M. Auffhammer, et al. 2008. *Climate change and Chicago. Projections and potential impacts.* Chicago: City of Chicago.

Hsiang, S., R. Kopp, A. Jina, M. Delgado, J. Rising, S. Mohan, et al. 2014. *American climate prospectus: Economic risks in the United States.* New York: Rhodium Group.

Hunt, A. 2017. "ECONADAPT toolbox." Accessed July 15, 2020. http://econadapt-toolbox.eu/.

ICE (Institution of Civil Engineers). 2020. *State of the Nation 2020: Infrastructure and the 2050 net-zero target.* London: ICE.

Kappes, M., B. Pozueta, K. Charles, M. Cayetano, and C. Rogelis. 2017. *Prioritizing climate resilient transport investments in a data-scarce environment: A practitioners' guide.* Global Facility for Disaster Reduction and Recovery. Washington, DC: World Bank.

Kotchen, M. J. 2011. "An economic framework for evaluating climate proofing investments on infrastructure." In Chap. 9 in *Paving the way for climate-resilient infrastructure: Guidance for practitioners and planners,* edited by C. Connelly, 106–115. New York: United Nations Development Program.

Lempert, R. J., J. R. Arnold, R. S. Pulwarty, K. Gordon, K. Greig, C. Hawkins-Hoffman, et al. 2018. "Reducing risks through adaptation actions." In Chap. 28 in Vol. 2 of *Impacts, risks, and adaptation in the United States: Fourth National Climate Assessment,* edited by D. R. Reidmiller, C. W. Avery, D. R. Easterling, K. E. Kunkel, K. L. M. Lewis, and T. K. Maycock, 1309–1345. Washington, DC: US Global Change Research Program.

Marcelo, D., C. Mandri-Perrott, S. House, and J. Schwartz. 2016. *Prioritizing infrastructure investment: A framework for government decision making.* Policy Research Working Paper 7674. Washington, DC: World Bank.

Markolf, S. A., M. V. Chester, D. A. Eisenberg, D. M. Iwaniec, C. I. Davidson, R. Zimmerman, et al. 2018. "Interdependent infrastructure as linked social, ecological, and technological systems (SETSs) to address lock-in and enhance resilience." *Earth's Future* 6 (12): 1638–1659.

McDaniels, T. L., S. E. Chang, D. Hawkins, G. Chew, and H. Longstaff. 2015. "Towards disaster-resilient cities: An approach for setting priorities in infrastructure mitigation efforts." *Environ. Syst. Decis.* 35 (2): 252–263.

McEvoy, D., P. Matczak, I. Banaszak, and A. Chorynski. 2010. "Framing adaptation to climate-related extreme events." *Mitigation Adapt. Strategies Global Change* 15 (7): 779–795.

Morss, R. E., O. V. Wilhelmi, G. A. Meehl, and L. Dilling. 2011. "Improving societal outcomes of extreme weather in a changing climate: An integrated perspective." *Annu. Rev. Environ. Resour.* 36 (1): 1–25.

Mountford, H., J. Corfee-Morlot, M. McGregor, F. Banajai, A. Bhattacharrya, J. Brand, et al. 2018. "Unlocking the inclusive growth story of the 21st century: Accelerating climate action in urgent times." Accessed March 20, 2019. https://newclimateeconomy. report/2018/wp-content/uploads/sites/6/2018/09/NCE_2018_FULL-REPORT.pdf.

Multihazard Mitigation Council. 2018. *Natural hazard mitigation saves: Utilities and transportation infrastructure* (P. Principal Investigator: Porter, K.; co-Principal Investigators: Scawthorn, C.; Huyck, C.; Investigators: Eguchi, R., Hu, Z.; Director: Schneider, Ed.). Washington, DC: Multihazard Mitigation Council. National Institute of Building Sciences.

NASEM (National Academies of Sciences Engineering and Medicine). 2020. *Incorporating the costs and benefits of adaptation measures in preparation for extreme weather events and climate change—Guidebook*. Washington, DC: Transportation Research Board.

NIAC (National Infrastructure Advisory Council). 2015. Executive Collaboration for the Nation's Strategic Critical Infrastructure Final Report and Recommendations. Accessed on November 1, 2018. https://www.cisa.gov/publication/niac-executive-collaboration-final-report.

Nikellis, A., and K. Sett. 2020. "Multihazard risk assessment and cost–benefit analysis of a bridge–roadway–levee system." *J. Struct. Eng.* 146 (5): 04020050.

Nordhaus, W. D. 1994. *Managing the global commons: The economics of climate change*. Cambridge, MA: MIT Press.

ORR (NYC Mayor's Office of Recovery and Resiliency). 2018. *Climate resiliency design guidelines*. 2nd ed. New York: ORR.

Pathways to Resilience (P2R) Partners. 2015. *Pathways to resilience: Transforming cities in a changing climate*. Oakland, CA: P2R Partners (Movement Strategy Center, Movement Generation, The Praxis Project, Reimagine! RP&E and the Kresge Foundation).

Pregnolato, M., A. Ford, V. Glenis, S. Wilkinson, and R. Dawson. 2017. "Impact of climate change on disruption to urban transport networks from pluvial flooding." *J. Infrastruct. Syst.* 23 (4): 04017015.

Republic of Ghana Ministry of Finance. 2016. *Climate change project prioritization tool and guidelines*. Accra, Ghana: Ministry of Finance.

Simmons, K. M., J. Czajkowski, and J. M. Done. 2018. "Economic effectiveness of implementing a statewide building code: The case of Florida." *Land Economics* 94 (2): 155–174.

Thaler, T., S. Fuchs, S. Priest, and N. Doorn. 2018. "Social justice in the context of adaptation to climate change—Reflecting on different policy approaches to distribute and allocate flood risk management." *Reg. Environ. Change* 18 (2): 305–309.

The Cities of Eugene and Springfield Oregon. 2014. "Regional Climate and Hazards Vulnerability Assessment." In *Eugene-Springfield Multi Jurisdictional Natural Hazards Mitigation Plan: 4-3–4-97*. Accessed September 3, 2019. https://www.eugene-or.gov/681/Emergency-Plans.

USACE (United States Army Corps of Engineers). 2019. *Watershed vulnerability assessment tool*. Washington, DC: USACE.

USGCRP (US Global Change Research Program). 2018. Vol. 2 of *Impacts, risks, and adaptation in the United States: The Fourth National Climate Assessment*, edited by D. R. Reidmiller, C. W. Avery, D. R. Easterling, K. E. Kunkel, K. L. M. Lewis, T. K. Maycock, et al. Washington, DC: US Government Publishing Office.

Wasley, E., K. Jacobs, J. Weiss, N. Preston, and M. Richmond. 2020. *Mapping climate exposure and climate information needs to water utility business functions*. Project 4729. Alexandria, VA: Water Research Foundation.

CHAPTER 5

Conclusions

Much of the existing infrastructure in the United States was designed for the climatic conditions of the twentieth century, with most current engineering minimum design standards not updated to account for recent and future climate change impacts (ASCE CACC 2018, Lopez-Cantu and Samaras 2018). In a recent report, ASCE (2019) identified six important trends that will affect the way in which civil engineers approach design and the built environment; climate change was identified as the greatest threat, whereas the others might present a mix of challenges and opportunities. In addition to the long-term changes in climatology and associated changes to normal operating conditions, increased frequency and intensity of extreme events are highlighted not only as a major threat to infrastructure but also as a catalyst to fund novel engineered solutions. In keeping with that report, this book examines the threats posed by future weather and climate extremes to our nation's infrastructure and provides an assessment of those infrastructure that will have the broadest benefits by preventing a cascade of negative impacts. The level of action already taken to prepare for climate change is also discussed and does vary, whereas example schemes are presented that provide suggestions of prioritization to mitigate short-term impacts and to prepare for long-term consequences within the country or a specific region of interest.

The increasing complexity and interconnectedness of society exacerbates its sensitivity to extreme weather and climate. Exposure to these hazards is also increasing, as a function of changes in the frequency of these events and as a result of urban growth in hazard-prone areas. Individual disasters can incur direct losses of billions of dollars (Munich Re 2018). Such losses could be dramatically reduced with appropriate adaptation measures. However, it is estimated that $20 trillion is needed over the next 10 years to bring US infrastructure up to minimum standards (ASCE 2017), and this figure does not include the shortfall needed to address the sustainability and resilience gap (Arcadis 2016). Thus, this book examines methodologies that may be used to develop a method of prioritization of adaptation measures that are needed to achieve the greatest benefits with limited time and financial resources.

The Fourth National Climate Assessment report (USGCRP 2018) states that the rate of global warming substantially outpaces natural climate variability and that the effects are evident in the frequency and intensity of extremes manifested through a myriad of meteorological and hydrological factors. Principal among these factors that will affect civil engineering infrastructure are precipitation extremes, temperature extremes, sea-level rise, tropical cyclones, and other extreme storm events. The current scientific consensus is that extreme precipitation will increase in intensity across the contiguous United States and be distributed as more frequent dry days punctuated by more frequent heavy rainfall events and flooding, with the moderate spectrum events in between decreasing overall in frequency (Wehner et al. 2017). Concomitant changes in precipitation and temperature will lead to an increase in the frequency and intensity of all drought types through changes in evaporation and decreased snowpack, with a likely parallel increase in the frequency, intensity, and areal extent of wildfires. Similarly, the combination of increased temperatures with the corresponding changes in precipitation frequency and type will result in changes in snowpack, as well as the magnitude and timing of spring snowmelt. Other examples of cascading events include postwildfire flooding and compounded storm surges and higher sea levels.

Such extremes as previously described are already affecting infrastructure resilience and durability, with increasingly adverse effects expected in many regions, and have potential ramifications for safety, environmental sustainability, economic vitality and mobility, and system reliability, particularly for vulnerable populations and urban infrastructure (Jacobs et al. 2018a). To look at the potential impacts of these extremes on various sector infrastructure types, the 16 infrastructure categories considered within the ASCE infrastructure report card (ASCE 2017) provided a starting point for this analysis. After combining overlapping sectors and omitting less critical sectors from the analysis, five overarching groups were considered: energy transmission, storage, and distribution; transportation (e.g., road and bridges; transit and aviation); drinking water and wastewater; flood protection infrastructure (e.g., levees and flood barrier; dams and reservoirs); and navigation/ports and harbors. Each sector was discussed in terms of the following:

1. Potential impacts of climate change,

2. Status of sector resiliency to future climate change,

3. Adaptation response and actions available to increase resiliency,

4. Interdependencies with other sectors that may facilitate cascading failures, and

5. Summary of the components of vulnerability for the sector.

The components of vulnerability were also summarized for each sector and as such were split into the potential impact (i.e., exposure + sensitivity) on the sector and its capacity to implement actions that will address potential future impacts of climate change. The authors' conclusions for each sector are summarized as follows:

1. Energy transmission, storage, and distribution: Because of the high exposure and sensitivity of the energy sector to a wide range of hydrological and meteorological impacts resulting from climate change and the strong

interdependencies between the energy sector and all other sectors, the energy sector is considered to be highly vulnerable to the potential effects of climate change. In particular, electricity generation will be affected by extreme temperatures, precipitation, and winds from extreme storms and will suffer reduced efficiency as temperatures rise. At the same time, the demand for electricity is expected to increase significantly in the future because of increased cooling demands, increased population, and a growing number of electric vehicles on the road. Because of its interdependency with all other critical infrastructure, all types of energy infrastructure are characterized as uniquely critical.

2. Transportation: Coupled with the aging and deteriorating infrastructure network associated with transportation, its location in flat low-lying areas, in general, heightens transportation exposure and sensitivity to weather extremes and changing climatic conditions, particularly in terms of flooding from extreme precipitation events and higher sea levels. Although some advancements have been made in understanding the risks posed to transportation infrastructure by climate change, these have mostly centered on vulnerability assessments. Furthermore, the studies focus on individual assets rather than the full network and the associated costs of replacement but not indirect costs from avoided disruptions. As with the energy sector, transportation is highly interlinked with other sectors, often providing critical support routes for evacuation and reconstruction activities. The lack of redundancy in all transportation forms makes this sector particularly vulnerable to unforeseen extreme weather events.

3. Drinking water and wastewater: Drinking water and wastewater systems are aging and increasing in fragility. Their location near rivers and coasts makes them highly exposed to flooding from fluvial, pluvial, and tidal sources and the attendant effects on water quality. Distribution and collection systems are sensitive to soil conditions and ground expansion or contraction because of fire, heat, desiccation, or saturation. Similarly, dam and reservoir operations are fairly responsive to precipitation variability under current river management practices, but water supply is sensitive to future changes in precipitation patterns and the adequacy of storage. Projected changes in precipitation and temperature are also likely to increase reservoir sedimentation and decrease water quality, as well as compromising dam structural integrity. Thus, although the adaptive capacity that has and continues to be built into systems, particularly those in larger urban areas, means that the current vulnerability is moderate, the fundamental importance of water and wastewater will result in severe impacts in the event of a loss.

4. Flood protection infrastructure: The majority of the nation's levees are more than 50 years old and are rated to be in *poor condition*, with many elements approaching the end of their service life. For example, in addition to changes in hydrometeorological conditions affecting the structural integrity of

earthen dams, the designed crest heights may no longer be adequate and are likely to be exceeded with increasing flood potential. Flood management is moving away from relying mainly on hard infrastructure, such as levees and dams, and is increasingly relying on a portfolio of nonstructural measures, thus increasing the adaptive capacity of this sector. However, the dependence on dams and levees to provide flood protection to other infrastructure systems increases the vulnerability of this sector.

5. Navigation/ports and harbors: The most concerning climate stressors for the navigation sector relate to forecasted changes in river flows and channel depths. River reaches with no locks or dams to maintain navigation depths are particularly vulnerable to low flows. Redundancy exists within the sector through the use of alternate shipping routes or road and rail freight; however, there is insufficient redundancy to accommodate the loss of barge capacity in the event of multiyear droughts. In contrast, port and harbor operations are most vulnerable to flooding, submersion by saline water, and sedimentation because of higher sea levels and higher and more frequent storm surges associated with an increased frequency and intensity of extreme coastal storm events. In addition, the vulnerability of port and harbor infrastructure is exacerbated by its strong dependence on supporting infrastructure such as energy and roads.

Following an analysis of the vulnerabilities of each sector to the impacts of climate change and their current state with respect to the implementation of solutions to address these vulnerabilities, a method of prioritization of projects is necessary, given the constrained resources. Such a prioritization will be unique to a particular country, region, or location depending on its exposure to specific manifestations of climate change (e.g., sea-level rise is not as much of a concern for inland sites) and its dependence on specific sectors (e.g., a region may have no reliance on inland navigation and ports and harbors). As such, the process of prioritization requires a step-by-step approach that initially ensures that an understanding of the decision context within which prioritization needs to occur is achieved; this includes an overall understanding of the climate change priorities and strategies already set in place through current policy and action plans along with a review of current and future projects within the sector of the project(s) under consideration. A list of potential technologies to be implemented and that address the specific objectives of the priorities and strategies identified needs to be developed, with each technology being further considered based on a number of factors, including costs and benefits, operation and maintenance, application potential and efficiency, and potential barriers to full implementation. To determine the extent to which a particular technology is appropriate, pertinent questions and criteria need to be devised, such as the following:

- What are the greatest sensitivities, exposures, consequences, and potential risk reductions to climate change indicators/impacts?
- Are there interdependencies and cascading/external impacts/benefits?

- What are the financial, economic, social, and environmental benefits and costs?

- Where is the greatest infrastructure investment gap, in terms of both location and sector?

- What are political/institutional barriers and how many agencies will be involved?

Based on these and other questions, a scoring method must be developed by which to interpret the answers to the questions/criteria, leading to project prioritization. Several examples of qualitative and quantitative methods were described and illustrated. The method adopted by any agency depends on the particular questions posed by the prioritization scheme and the audience to which this information is directed. Finally, the uncertainties associated with the results from the prioritization scheme, particularly if they are more quantitative in nature, must be taken into consideration.

Even after a prioritization scheme is developed, considerable challenges remain regarding the implementation of the highest priority adaptation actions, not the least of which is the financial burden of improved construction. Justification for project expenditure often relies on a cost–benefit analysis. However, there are limited data available to assess the economic effects of climate change and, hence, the benefits of construction now rather than at some point in the future when the effects are more apparent. In addition to the financial burden, the other challenges and gaps that must be addressed include the following:

1. Accelerating the resilience of infrastructure to current and future weather and climate extremes cannot be achieved by a single discipline (de Bruijn et al. 2017); the current lack of resilience (i.e. fragility) stems from a traditional design approach that considers only one event, discipline, or infrastructure component at a time (Markolf et al. 2018). The complexity of problems faced by an increasingly urbanized society, changing climate conditions, and economic constraints can be addressed only by creative thinking informed by a technically and socially diverse group (Conner 2016, Douglas et al. 2017, Hunt et al. 2015).

2. Reducing the disconnect between research and practical applications (Cash et al. 2006), such as how to incorporate future projections and scientific uncertainties into the decision-making frameworks that govern the design and engineering of infrastructure assets and systems (NASEM 2018). National, subnational, and local governments are being asked to increase the resilience of aging infrastructure systems with declining budgets (Jacobs et al. 2018b). Similarly, these actors are increasingly being asked to design infrastructure systems to address interdependencies between infrastructure systems and climate impacts that cross multiple sectors (Jacobs et al. 2018a).

3. Understanding the implications of climate change on infrastructure design needs to be introduced at the undergraduate level, using a transdisciplinary approach to ensure that graduates are able to deliver the benefits that

society expects (Gauvreau 2018). Continued exposure to multidisciplinary knowledge beyond graduation to develop creative out-of-the-box thinking is critical for surmounting the daily design challenges balancing economic, environmental, political, and other considerations (Idi et al. 2018). Thus, early career researchers and practitioners need training to understand the impacts of climate change, to analyze risks, to conduct research, and to plan, design, construct, and maintain infrastructure using appropriate methods (Cooper et al. 2019).

4. New research on climate resilient materials, planning, design, and construction processes for future operational and extreme weather conditions is critical (Pacheco-Torgal 2014). However, a considerable barrier is the shortage of researchers and practitioners with the capability to work at the boundaries of several disciplines and to advance this work.

5. Research and development is required to improve forecasts of impact severity and location to improve planning and prioritization. Such research relies not only on improved climate models but also on improved integration between knowledge users and producers (Morss et al. 2018) to determine where the focus of research should be directed. Cross-sectoral collaboration and communication will also enhance the development of data that are useful and usable by decision-makers and has been shown to enhance resilience (Rodina 2019).

To address many of these challenges, an interdisciplinary approach involving climate scientists, engineers, economists, policymakers, and other experts is needed to achieve the dramatic changes needed in the planning, design, construction, maintenance, and decommissioning processes. Collaborative research is needed to understand how climate change will impact specific locations, regions, and assets over different time scales. Improved two-way communication will be essential to address uncertainties in climate modeling and integrate these considerations into engineering and design. Finally, the policies and laws that inform infrastructure decision-making and investment, including the standards that have been put in place by ASCE, require a thorough review to integrate climate change considerations into decision-making and design processes more effectively. As demonstrated by some recent projects, this process will likely combine *bottom-up* and *top-down* approaches such as community-led risk assessments and enhanced requests for proposals or financing arrangements that are contingent on adaptation planning.

References

Arcadis. 2016. "Global built asset wealth index 2015." Accessed June 9, 2019. https://www.arcadis.com/en/global/our-perspectives/global-built-asset-wealth-index/.

ASCE. 2017. *Infrastructure report card. A comprehensive assessment of America's infrastructure.* Reston, VA: ASCE.

ASCE. 2019. *Future world vision: Infrastructure reimagined.* Reston, VA: ASCE.

ASCE CACC (Committee on Adaptation to a Changing Climate). 2018. *Climate-resilient infrastructure*, edited by B. M. Ayyub. Reston, VA: ASCE.

Cash, D. W., J. C. Borck, and A. G. Patt. 2006. "Countering the loading-dock approach to linking science and decision making: Comparative analysis of EI Niño/Southern Oscillation (ENSO) forecasting systems." *Sci. Technol. Hum. Values* 31 (4): 465–494.

Conner, T. W. 2016. "Representation and collaboration: Exploring the role of shared identity in the collaborative process." *Public Admin. Rev.* 76 (2): 288–301.

Cooper, D., R. Springer, J. Benner, D. Bloom, E. Carrillo, A. Carroll, et al. 2019. *Supporting the changing research practices of civil and environmental engineering scholars.* New York: Ithaka S+R.

de Bruijn, K., J. Buurman, M. Mens, R. Dahm, and F. Klijn. 2017. "Resilience in practice: Five principles to enable societies to cope with extreme weather events." *Environ. Sci. Policy* 70: 21–30.

Douglas, E., J. Jacobs, K. Hayhoe, L. Silka, J. Daniel, M. Collins, et al. 2017. "Progress and challenges in incorporating climate change information into transportation research and design." *J. Infrastruct. Syst.* 23 (4): 04017018.

Gauvreau, P. 2018. "Sustainable education for bridge engineers." *J. Traffic Transp. Eng.* 5 (6): 510–519.

Hunt, V., D. Layton, and S. Prince. 2015. "Diversity matters." Accessed March 1, 2020. https://www.mckinsey.com/~/media/mckinsey/businessfunctions/organization/ourinsights/whydiversitymatters/diversitymatters.ashx.

Idi, D. B., and K. A. M. Khaidzir. 2018. "Critical perspective of design collaboration: A review." *Front. Archit. Res.* 7 (4): 544–560. https://doi.org/10.1016/j.foar.2018.10.002.

Jacobs, J. M., L. R. Cattaneo, W. Sweet, and T. Mansfield. 2018b. "Recent and future outlooks for nuisance flooding impacts on roadways on the US East Coast." *Transp. Res. Rec.* 2672 (2): 1–10.

Jacobs, J. M., M. Culp, L. Cattaneo, P. Chinowsky, A. Choate, S. DesRoches, et al. 2018a. "Transportation." In Vol. 2 of *Impacts, risks, and adaptation in the United States: Fourth National Climate Assessment*, edited by D. R. Reidmiller, C. W. Avery, D. R. Easterling, K. E. Kunkel, K. L. M. Lewis, T. K. Maycock, et al., 479–511. Washington, DC: US Global Change Research Program.

Lopez-Cantu, T., and C. Samaras. 2018. "Temporal and spatial evaluation of stormwater engineering standards reveals risks and priorities across the United States." *Environ. Res. Lett.* 13 (7): 074006.

Markolf, S. A., M. V. Chester, D. A. Eisenberg, D. M. Iwaniec, C. I. Davidson, R. Zimmerman, et al. 2018. "Interdependent infrastructure as linked social, ecological, and technological systems (SETSs) to address lock-in and enhance resilience." *Earth's Future* 6 (12): 1638–1659.

Morss, R. E., J. M. Done, H. Lazrus, E. Towler, and M. R. Tye. 2018. "Connecting predictive capacity to stakeholder needs." *US CLIVAR Var.* 16 (3): 24–30.

Munich Re. 2018. "NatCatSERVICE: Loss events worldwide 2017. Geographical overview." Accessed March 6, 2018. https://www.munichre.com/site/corporate/get/params_E976667823_Dattachment/1627370/MunichRe-NatCat-2017-World-Map.pdf.

NASEM (National Academies of Sciences, Engineering, and Medicine). 2018. *Critical issues in transportation 2019.* Washington, DC: National Academies Press.

Pacheco-Torgal, F. 2014. "Eco-efficient construction and building materials research under the EU framework programme horizon 2020." *Constr. Build. Mater.* 51: 151–162.

Rodina, L. 2019. "Defining 'water resilience': Debates, concepts, approaches, and gaps." *WIREs Water* 6 (2): e1334.

USGCRP (US Global Change Research Program). 2018. *Impacts, risks, and adaptation in the United States: Fourth National Climate Assessment, Volume II*, edited by D. R. Reidmiller, C. W. Avery, D. R. Easterling, K. E. Kunkel, K. L. M. Lewis, T. K. Maycock, et al. Washington, DC: USGCRP.

Wehner, M. F., J. R. Arnold, T. Knutson, K. E. Kunkel, and A. N. LeGrande. 2017. "Droughts, floods, and wildfires." In Vol. 1 of *Climate science special report: Fourth National Climate Assessment*, edited by D. J. Wuebbles, D. W. Fahey, K. A. Hibbard, D. J. Dokken, B. C. Stewart, and T. K. Maycock, 231–256. Washington, DC: USGCRP.

Glossary

Adaptive Capacity: Ability to absorb gradual changes in operating conditions; flexibility to extend or modify a structural component to meet new conditions; or presence of dynamic adaptation planning to reassess and adjust with time.

Cascading Failure: Failure or series of failures arising in response to the loss of service of another infrastructure element or system.

Disruptive Events: Events that are rare, or extreme, or that have high consequences. They are classified as extreme, high impact, rare, and severe (Stephenson 2008).

> **Extreme**: Those with very high (or low) values of certain meteorological parameters. In combination with statistical approaches, these are often defined as temperatures, precipitation, wind speeds, and so on that have a low annual probability of occurrence, also referred to as target design thresholds such as the top or bottom 1% of the observed probability density function.

> **High Impact**: Both short-term severe weather systems, such as intense precipitation, or long-duration events, such as prolonged droughts, which lead to high human cost.

> **Rare**: Those that have a low probability of occurrence. These events are often rarer than those codified for design purposes and can lead to considerable damage and devastation because society, infrastructure, and ecosystem are not well adapted to absorb these events. Where codified for engineering purposes, rare events represent an *acceptable level of failure* such as a 0.01% annual probability of occurrence. With a changing climate, we are beginning to experience such rare events with greater frequency and need to incorporate their occurrence in adaptation plans.

> **Severe**: Measured as a function of both the probability of the hazard or event and the consequences (number of people exposed and their vulnerability). Severity is often measured in terms of the long-term losses in, for example, lives, financial capital, or ecosystems.

Exposure: Presence of infrastructure, services, and resources, people and livelihoods, or economic, social, or cultural assets in places that could be adversely affected by a disruptive event (IPCC 2012).

Fragility: Probability of failure as a result of one or more changes in exposure or susceptibility to a disruptive event; inadequate preparation for and adaptation to changing conditions; and lack of capacity to absorb persistent threats, changes, and uncertainty.

Hazard: Source of potential harm or a condition that may result from an external cause (e.g., earthquake, flood, or human agency) or an internal vulnerability with the potential to initiate a failure mode (ASCE CACC 2018). In this report, the term hazard specifically refers to weather- or climate-related physical events or their physical impacts.

Impact: Effect on natural and human systems, primarily from a disruptive event. Impact is a function of the probability of a disruptive event (i.e., exposure), the likely response to that event, and the degree of importance of the infrastructure, services, and resources or economic, social, or cultural assets.

Resilience: Capacity to withstand and rapidly recover from a disruptive event, responding or reorganizing in ways that maintain their essential function or purpose by embracing the uncertainties of current and future climate conditions through redundancy and flexible design thresholds.

Risk: Probability of a hazard occurring multiplied by the impacts if the disruptive events occur, giving the potential for adverse consequences to arise.

Sensitivity: Potential degree of damage directly or indirectly inflicted on infrastructure, services, and resources or economic, social, or cultural assets by a disruptive event. Damage can be measured in physical units encompassing the total or partial destruction of assets within the exposed area or as monetary replacement costs. Indirect damages are those arising from a cascading failure (e.g., a flooded highway preventing access to repair a damaged power substation, leading to power failure to critical facilities such as hospitals).

Status: Degree to which a sector is showing interest or sophistication in understanding the risks posed by climate hazards, the development of adaptation plans, and implementation of those plans or other resilience measures.

Vulnerability: Degree to which infrastructure is susceptible to, or unable to cope with the effects of a disruptive event, and the propensity to be adversely affected by climate change.

References

ASCE CACC (Committee on Adaptation to a Changing Climate). 2018. *Climate-resilient infrastructure*, edited by B. M. Ayyub. Reston, VA: ASCE.

IPCC (Intergovernmental Panel on Climate Change). 2012. "Figure SPM.1 from IPCC, 2012: 'Summary for policymakers.'" In *Managing the risks of extreme events and disasters to advance climate change adaptation*, edited by C. B. Field, V. Barros, T. F. Stocker, D. Qin, D. J. Dokken, K. L. Ebi, et al., 3–21. A Special Report of Working Groups I and II of the Intergovernmental Panel on Climate Change. Cambridge, UK: Cambridge University Press.

Index

Note: Page numbers followed by *f* and *t* indicate figures and tables.